Introduction to Peptide Science

Introduction to Peptide Science

IAN W. HAMLEY
School of Chemistry, University of Reading, UK

This edition first published 2020
© 2020 John Wiley & Sons Ltd

The right of Ian W. Hamley to be identified as the author of this work has been asserted in accordance with law.

Registered Offices
John Wiley & Sons, Inc., 111 River Street, Hoboken, NJ 07030, USA
John Wiley & Sons Ltd, The Atrium, Southern Gate, Chichester, West Sussex, PO19 8SQ, UK

Editorial Office
The Atrium, Southern Gate, Chichester, West Sussex, PO19 8SQ, UK

For details of our global editorial offices, customer services, and more information about Wiley products visit us at www.wiley.com.

Wiley also publishes its books in a variety of electronic formats and by print-on-demand. Some content that appears in standard print versions of this book may not be available in other formats.

Library of Congress Cataloging-in-Publication Data

Names: Hamley, Ian W., author.
Title: Introduction to peptide science / Ian W. Hamley.
Description: First edition. | Hoboken, NJ : Wiley, 2020. | Includes
    bibliographical references and index.
Identifiers: LCCN 2020013505 (print) | LCCN 2020013506 (ebook) | ISBN
    9781119698173 (paperback) | ISBN 9781119698180 (adobe pdf) | ISBN
    9781119698197 (epub)
Subjects: MESH: Peptides | Antimicrobial Cationic Peptides | Peptide
    Hormones | Nanotubes, Peptide
Classification: LCC QP552.P4 (print) | LCC QP552.P4 (ebook) | NLM QU 68
    | DDC 612/.015756–dc23
LC record available at https://lccn.loc.gov/2020013505
LC ebook record available at https://lccn.loc.gov/2020013506

Cover Design: Wiley
Cover Image: © MOLEKUUL/Getty Images

Set in 10/13pt SabonLTStd by SPi Global, Chennai, India

Printed and bound by CPI Group (UK) Ltd, Croydon, CR0 4YY

10 9 8 7 6 5 4 3 2 1

Contents

Preface

Welcome (or welcome back!) to the wonderful world of peptide science. Peptides are fascinating chain molecules, built from amino acids, that nature has evolved to fulfil an incredible range of structural and functional roles. Of particular importance are peptide hormones, a major class of signalling molecules in the body. But there are many other essential peptides ensuring your body keeps working. Peptides crop up everywhere, along with their larger cousins proteins, as components of biological structures in silk, collagen, amyloid, and many other biomaterials. Amyloid is also now implicated as a 'pathological' agent in many diseases that are becoming of increasing concern, including neurodegenerative diseases such as Alzheimer's. Peptide hormones are playing an important role in conditions associated with a modern lifestyle, such as diabetes and the related condition of obesity.

Nature uses an alphabet of 20 natural amino acids to build peptides. Peptides have been synthesized in the laboratory for decades and in the last 50 years automated synthesis methods have been introduced, been rapidly developed, and have become established. Scientists are now designing original sequences and incorporating novel residues, functionalities, and configurations into peptides and are also creating conjugates of peptides with lipids, glyco-saccharides, and polymers.

This book covers the basic properties of peptides, looking at essential synthesis methods and peptide aggregate structures, including amyloid and other nanostructures such as peptide nanotubes and peptide gels. Medically related applications are considered in the final two chapters, devoted to antimicrobial peptides and peptide therapeutics. These are discussed in the context of peptide hormones, from which many of the most important peptide therapeutics introduced into practice (so far) are derived.

This book is intended to provide a broad coverage of peptide science that is otherwise lacking in existing textbooks. There are several excellent books that cover peptide synthesis in detail and a few books cover basic properties, with the best coverage often being in introductory texts on proteins or general biochemistry. There are also specialist texts on peptide therapeutics and antimicrobial peptides, and a very few on amyloid. However, I felt there was a need for a compact introductory text that covers applications such as therapeutics and biomaterials. This book aims to achieve that. In addition, it covers at an introductory level a number of modern developments in the field that are not included in older texts in the areas of synthesis and aggregation/self-assembly. The book includes aspects of synthetic chemistry, biochemistry, and biophysics relevant to understanding peptide science. This is an interdisciplinary field, so these are not considered to be exclusive terms, nor is this an exhaustive list of disciplines that feed into or from peptide science.

I have intended this book to be an introduction for senior undergraduates of chemistry, biochemistry, or biology and so should be useful as a supplementary text for junior courses in these and allied subjects. In addition it contains, in a compact and easy to reach form, much material that should be a valuable reference source for researchers in the field. All the texts I consulted in the course of writing the book, including existing books in the field along with key articles (mainly review articles) that I referred to, are cited in the Bibliography sections at the end of each chapter. Of course these do not provide a complete reference list on the subject; it would be impossible to compile such a list, given the huge volume of research in this exciting and fast-moving field.

I apologize in advance for any errors or omissions and would be grateful to be informed of these. I would also be happy to receive any other feedback on the book as this would be very useful if a future edition emerges. At the moment, I will take a justified break from peptide book writing, although I've enjoyed the process and have learnt many useful new things. I hope you enjoy it in a similar style.

I would like to acknowledge my editor Jenny Cossham for supporting this project and my group of great students and postdocs who have helped immensely as we have learned together over the last couple of decades about peptides and their applications. Also thanks to all my many valued collaborators over the years, from around the world. Finally, I am very grateful to my family for their extracurricular support, and in the case of my wife Valeria for curricular support as well!

Ian W. Hamley
University of Reading, UK, 2020

1

Basic Properties

1.1 INTRODUCTION

Peptides and proteins are essential biological molecules and assemblies, and many structures and functions of organisms are derived by using them. Peptides and proteins comprise chains of amino acids and are produced from DNA in the ribosome via messenger RNA through three-letter codons, i.e. each amino acid is represented by a triplet of nucleotides. Peptides are also known as 'polypeptides' to signify that they are polymers formed of amino acids. In fact, short peptides as considered in this book have the properties of oligomers rather than polymers.

Although there is no rigorous definition of the difference between peptides and proteins in this book, a typical peptide would have up to 100 residues, and longer chains would be considered proteins. In fact, standard synthesis methods lead to peptides up to 50–60 residues in length, which may also be considered a cut-off, with somewhat larger chains being described as 'mini-proteins'.

This chapter is organized as follows. In Section 1.2, the essential properties of the 20 standard natural amino acids that peptides and proteins are built from are first considered, including an analysis of reactivity, charge, and hydrophobicity. When incorporated in a peptide or protein, the term 'residue' is used for the group formed from the corresponding amino acid. Non-natural residues that are found in a few natural and synthetic peptides are also introduced in Section 1.2, especially those which are mentioned in peptide sequences elsewhere in this book. Then, in Section 1.3, the nature and geometry of the peptide bond are considered. The main secondary structures adopted by peptides are discussed in Section 1.4. There follows

Introduction to Peptide Science, First Edition. Ian W. Hamley.
© 2020 John Wiley & Sons Ltd. Published 2020 by John Wiley & Sons Ltd.

(Section 1.5) a description of characterization methods used to determine peptide structure and conformation. The chapter concludes with a list of useful peptide websites and selected software, including databases and property calculators.

1.2   PROPERTIES OF AMINO ACIDS

Amino acids are chiral molecules, meaning that there is a distinct spatial arrangement of substituents around the central backbone carbon atom, which is termed a C_α atom. This is a stereogenic centre. There are two possible arrangements with different 'handedness', which are mirror images. These forms are termed enantiomers. There are two forms, termed L- and D-amino acids (note the use of a small capital letter). In the L form of a residue in a peptide, the substituents are arranged as shown in Figure 1.1, which can be remembered via the CORN mnemonic, with the CO, R, and N groups in a clockwise configuration from left, when viewed along the bond to the H atom. In nature, peptide and protein structures are almost entirely built from L-amino acids. Molecules that contain amino acids may have more than one stereocentre; for example, Figure 1.2 shows the diastereomers of threonine, L- and D-threonine, in the form of Fischer projections. Figure 1.2 also includes the two *allo-* forms, which are rarely found in natural peptides (isoleucine also has four diastereomers). The *R,S* notation for the stereocentres according to the Cahn–Ingold–Prelog rules are also shown. Fischer projections are planar representations of an enantiomer (named after Emil Fischer) in which bonds denoted as horizontal extend above the plane of the paper and vertical bonds denote those which extend below the plane. In contrast to (mono)stereoisomers, diastereoisomers can have distinct chemical and physical properties (for example melting points).

The peptide bond has partial double bond character, so its length 1.33 Å (0.133 nm) is significantly shorter than a usual C–N bond length of 1.45 Å

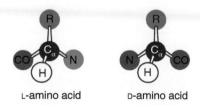

L-amino acid   D-amino acid

Figure 1.1   Enantiomers of amino acid residues in peptides. Left: L-amino acid, right: D-amino acid.

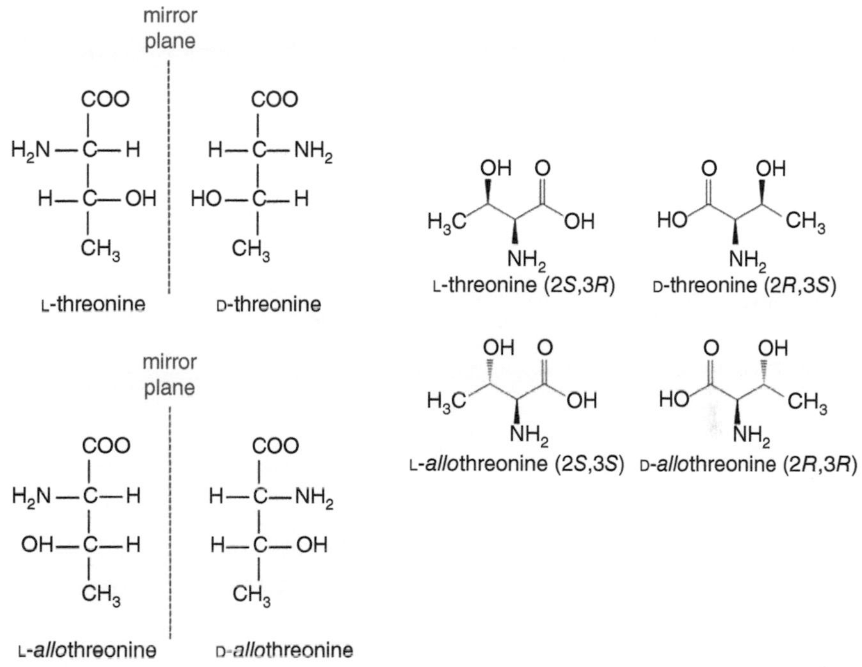

Figure 1.2 Fischer projections and structures of L-threonine, D-threonine, L-*allo*threonine, and D-*allo*threonine.

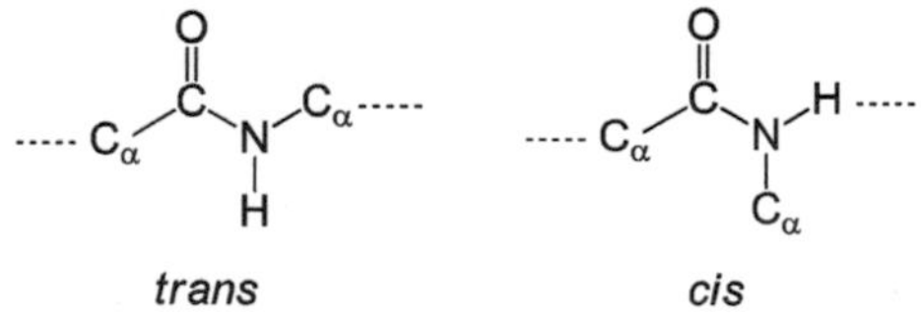

Figure 1.3 Configurations of a peptide bond.

(0.145 nm). The partial double bond character leads to restricted rotation and a preferred co-planar arrangement. Two configurations of the planar peptide bond are possible, called *trans* and *cis*, shown in Figure 1.3. The *trans* form is favoured energetically.

The 20 standard (so-called canonical) amino acids found in nature with different side chains are shown in Figure 1.4. They are grouped according to amino acid polarity or charge and, for the non-polar amino acids, according to whether they have an aliphatic or aromatic substituent. All the amino acids except glycine are chiral, being present as either L- or D-enantiomers. In nature, L-enantiomers are the standard form, although D-amino acids are

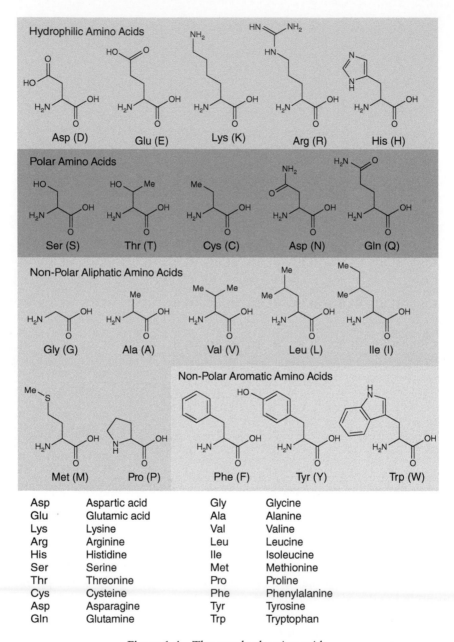

Asp — Aspartic acid
Glu — Glutamic acid
Lys — Lysine
Arg — Arginine
His — Histidine
Ser — Serine
Thr — Threonine
Cys — Cysteine
Asp — Asparagine
Gln — Glutamine

Gly — Glycine
Ala — Alanine
Val — Valine
Leu — Leucine
Ile — Isoleucine
Met — Methionine
Pro — Proline
Phe — Phenylalanine
Tyr — Tyrosine
Trp — Tryptophan

Figure 1.4   The standard amino acids.

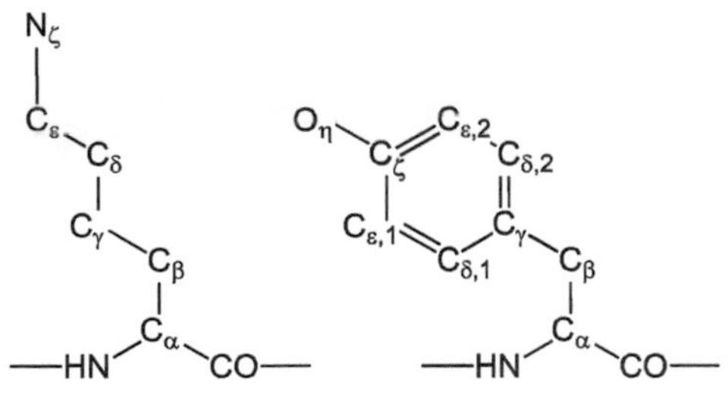

Figure 1.5 Amino acid side chain labelling nomenclature.

present in some natural peptides. Figure 1.4 does not show the enantiomeric forms. Here, and elsewhere, the symbol Me in a chemical formula indicates methyl (CH_3).

The labelling scheme for (non-hydrogen) atoms in amino acid side chains is shown in Figure 1.5, as exemplified for lysine and tyrosine.

The hydrophilic amino acids comprise two that are normally positively charged (except at low pH, see Table 1.2 and associated discussion) – these are the anionic residues Asp (D) and Glu (E). These differ only by the addition of an extra methylene group in the side chain of E. Although the pK_a values of the side chain residues are similar, the length of the side chain (and so the location of the charge) does significantly influence the conformation of the backbone and the reactivity of these groups. Section 2.11.5 discusses some reactions using these residues.

The two basic residues, Lys (K) and Arg (R), are cationic under most conditions, except high pH. Lysine is used for many conjugation reactions, as discussed in Sections 2.11.2 and 2.11.5. Arginine contains a strongly basic guanidino group, which has a resonance structure shown for arginine in the charged state in Figure 1.6, and among its properties it is able to form bidentate hydrogen bonds (see Figure 4.10).

Histidine can also exhibit basic character, although its pK_a is within the range of physiological pH values. It is unique among amino acids in having an imidazole group, and in the charged state the positive charge is shared between the nitrogen atoms by resonance. There are also two tautomeric forms in the uncharged state, depending on which nitrogen has an attached hydrogen atom.

Serine and threonine are related polar amino acids containing hydroxyl groups. They differ only by the addition of a methyl group in serine. Both these residues improve the solubility of peptides. Serine is important in the activity of many enzymes, such as serine proteases. There are two diastereomers of threonine, as discussed above (Figure 1.2).

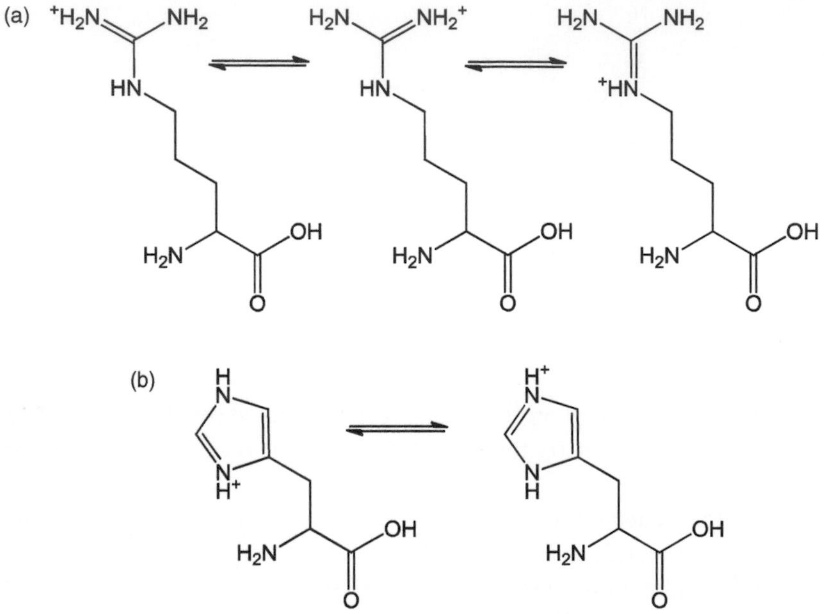

Figure 1.6 Resonance structures of (a) arginine and (b) histidine.

Cysteine is one of the two residues that contain sulfur in the thiol (sulfhydryl) group. This is very commonly used as a tag in bioconjugation reactions, as described further in Sections 2.11.2 and 2.11.5. The thiol group in cysteine can also form disulfide links, which are common stabilizing elements in proteins, and are also used in natural and synthetic cyclic peptides. Disulfide bridges are discussed further in Section 1.4 and the synthesis of cyclic peptides including disulfide bridging is discussed in Section 2.6. The sulfhydryl group ionizes at high pH to give a $-CH_2-S^-$ side chain.

The two residues Asn (N) and Gln (Q) with side chain amide groups differ only in side chain length. They are the amide forms of the acidic residues Asp (D) and Glu (E) respectively. These residues are polar, but do not ionize. The amide groups act as both hydrogen donor and hydrogen acceptor. The amide groups are labile at very high or low pH values and high temperatures and these residues can undergo deamidation to form Asp and Glu. N-Terminal glutamine residues can spontaneously cyclize to form pyroglutamic acid, which is shown in Table 1.3.

Glycine is the simplest amino acid, and the only one that is non-chiral because it only has a second hydrogen atom as its 'side group'. The amino acids Ala (A), Val (V), Leu (L), and Ile (I) constitute hydrophobic aliphatic residues. The hydrophobicity of these residues increases with the side-chain size. These residues are inert although they tend to associate through hydrophobic interactions, this being important in the stabilization of β-sheet structures for example. Ile contains an additional asymmetric centre (C_β) but only one diastereomer (L-isoleucine) occurs naturally. Methionine is grouped with the non-polar aliphatic residues in Figure 1.4, although because it contains a sulfur atom it is related to cysteine. However, the thiol group in cysteine is methylated, leading to a non-polar character and a lack of ionizability. The sulfur atom in methionine is susceptible to oxidation, forming first a sulfoxide, then a sulfone. Proline is a unique residue in that it does not contain an amide hydrogen able to participate in hydrogen bonds. In addition, the ring in proline leads to a conformational constraint that increases the fraction of *cis-* configured peptide chain before the Pro residue.

The three aromatic amino acids, Phe (F), Tyr (Y), and Trp (W), are hydrophobic. Phe and Trp are not very reactive under most conditions. The –OH group in Tyr is reactive and ionizes at high pH, under which conditions dityrosine formation is possible. The indole group in Trp leads to the intrinsic fluorescence of this residue, which is a useful property in aggregation assays since the fluorescence is sensitive to the local microenvironment. The aromatic groups in Phe, Tyr, and Trp can lead to π-stacking interactions which can stabilize some secondary and aggregate structures (Section 1.4). In addition, the aromatic groups have characteristic peaks in their UV spectra, which can be used in peptide/protein concentration measurements (Section 1.5.1).

Important properties of the amino acids, including the pK_a values of the side chains as well as hydrophobicity, are listed in Table 1.1. Other properties of amino acids are included, such as their typical abundance in proteins (from database analysis), van der Waals volumes, and hydrophobicity. There are a considerable number of published hydrophobicity scales, examples of the most well-known ones are included in Table 1.1. These are obtained from measurements of free energies of partition. Kyte and Doolittle presented a hydropathy scale based on water-vapour transfer free energy and the tendency of amino acids to be on the exterior or interior of proteins. This scale correlates to a good degree to the water-vapour ΔG(transfer). White and Wimley present tables of similarly determined ΔG(partition) values, ΔG(water/lipid interface), and ΔG(water/octanol). The trend in the Wimley–White ΔG values is clear when plotted (Figure 1.7). Other hydrophobicity scales have been proposed.

Since many peptides are charged due to the presence of residues with charged side chains and/or charged termini, analysis of pK_a values enables

Table 1.1 Properties of amino acids.

Symbol (one letter code)	Abundance in proteins (%)	Side chain pK_a	van der Waals volume (Å^3)	Hydrophobicity: Kyte–Doolittle hydropathy index	Hydrophobicity: Wimley–White interface scale, ΔG (kcal mol^{-1})	Hydrophobicity: Wimley–White octanol scale, ΔG (kcal mol^{-1})
Gly (G)	7.2		48	−0.4	0.01	1.15
Ala (A)	7.8		67	1.8	0.17	0.50
Val (V)	6.6		105	4.2	0.07	−0.46
Leu (L)	9.1		124	3.8	−0.56	−1.25
Ile (I)	5.3		124	4.5	−0.31	−1.12
Met (M)	2.2		124	1.9	−0.23	−0.67
Pro (P)	5.2		90	−1.6	0.45	0.14
Phe (F)	3.9		135	2.8	−1.13	−1.71
Tyr (Y)	3.2	10.5	141	−1.3	−0.94	−0.71
Trp (W)	1.4		163	−0.9	−1.85	−2.09
Ser (S)	6.8		73	−0.8	0.13	0.46
Thr (T)	5.9		93	−0.7	0.14	0.25
Cys (C)	1.9	8.4	86	2.5	−0.24	−0.02
Asn (N)	4.3		96	−3.5	0.42	0.85
Gln (Q)	4.3		114	−3.5	0.58	0.77
Asp (D)	5.3	3.9	91	−3.5	1.23	3.64
Glu (E)	6.3	4.0	109	−3.5	2.02	3.63
Lys (K)	5.9	10.5	135	−3.9	0.99	2.80
Arg (R)	5.1	12.5	148	−4.5	0.81	1.81
His (H)	2.3	6.0	118	−3.2	0.96	2.33

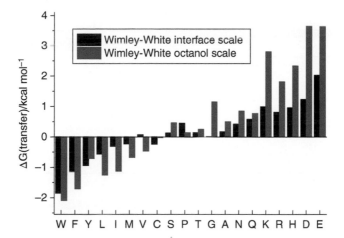

Figure 1.7 Hydrophobicity of amino acids, according to the Wimley–White scale. The hydrophobicity decreases left-to-right.

estimation of the overall net charge on the peptides (see Eq. (1.4) below). The isoelectric point (pI) is another useful characteristic of peptides; this corresponds to the pH at which the net charge is zero. Table 1.2 summarizes the state of charged amino acids below and above the respective side chain pK_a values. For the termini, typical pK_a values are $pK_a = 9$–10 for the N-terminal NH_3 and $pK_a = 2$–3 for the C-terminal COOH. It can be assumed that when the solution $pH = pK_a$, equal numbers of the charged and uncharged species will be present. It should be noted that pK_a values of residues in peptides (such as those quoted in Tables 1.1 and 1.2) are typical values; the pK_a of a particular residue will depend on its local environment (for example it will be modified by the presence of other nearby charged residues). Acid or base titrations may be used to experimentally determine pK_a values for residues in short charged peptides. Figure 1.8 shows a schematic titration curve.

Close to the pK_a, the two forms of the peptide can be represented as an acid–base equilibrium. Using the Henderson–Hasselbalch equation, the charge can be obtained.

$$HA \rightleftharpoons H^+ + A^-$$

The acid dissociation constant is given by

$$K_a = \frac{[A^-][H^+]}{[HA]} \tag{1.1}$$

The Henderson–Hasselbalch equation is

$$pH = pK_a + \log_{10}\left(\frac{[A^-]}{[HA]}\right) \tag{1.2}$$

Table 1.2 Side chain protonation depending on pH.

Amino acid	pH < pK_a	pK_a	pH > pK_a
Asp (D)	–COOH	3.9	–COO⁻
Glu (E)	–COOH	4.0	–COO⁻
Lys	–NH$_3^+$	10.5	–NH$_2$
Arg (R)	–C(NH$_2$)$_2^+$	12.5	–C(NH$_2$)(NH)
His (H)	imH⁺	6.0	imH
Tyr (Y)	–PheOH	10.5	–PheO⁻
Cys (C)	–SH	8.4	–S⁻

im = imidazole.

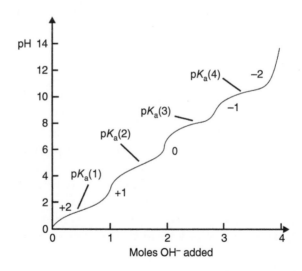

Figure 1.8 Schematic titration curve for a tetrapeptide NH$_2$–EGAK–COOH, plotting pH versus molar concentration of OH⁻ per mol of tetrapeptide. The peptide net charge is indicated between pK_a values (defined as the mid-point of the pseudo-plateaus) as shown. Here, $pK_a(1)$ is that for the C terminus, $pK_a(2)$ is for the E residue, $pK_a(3)$ is for the N terminus and $pK_a(4)$ is for the K residue. Net charge +2 corresponds to NH$_3^+$–(E)GA(K⁺)–COOH, +1 to NH$_3^+$–(E)GA(K⁺)–COO⁻, 0 to NH$_3^+$–(E⁻)GA(K⁺)–COO⁻, –1 to NH$_2$–(E⁻)GA(K⁺)–COO⁻, and –2 to NH$_2$–(E⁻)GA(K)–COO⁻, where E⁻ denotes charged glutamic acid (with COO⁻ side chain terminus) and K⁺ denotes charged lysine with NH$_3^+$ side chain terminus. The pI of this peptide is at pH = 7.

So the fractional charge

$$\frac{[A^-]}{[A^-] + [HA]} = \frac{10^{(pH-pK_a)}}{1 + 10^{(pH-pK_a)}} \tag{1.3}$$

For example, for lysine at pH 10, using the side chain $pK_a = 10.5$ (Table 1.2),

$$\frac{[A^-]}{[A^-] + [HA]} = \frac{10^{(pH-pK_a)}}{1 + 10^{(pH-pK_a)}} = \frac{10^{-0.5}}{1 + 10^{-0.5}} = 0.24$$

So approximately one quarter of the lysine residues are uncharged (in the form $[A^-]$, i.e. NH_2) under these conditions, which means for example for a peptide containing four charged lysine residues, three of them will be charged. The same result can be obtained from the formula for the net charge, q:

$$q = \sum_i N_i \frac{+1}{1 + 10^{+(pH-pK_{a,i})}} + \sum_j N_j \frac{-1}{1 + 10^{-(pH-pK_{a,j})}} \qquad (1.4)$$

where the index 'i' labels basic residues (total N_i of each type with $pK_{a,i}$) and 'j' labels acidic residues (N_j of each type with $pK_{a,j}$).

Several websites (listed in Table 1.6) provide convenient calculators for peptide net charge, pI, and other properties (molar mass) based on input sequences.

The amphiphilicity of α-helices can be quantified via the hydrophobic moment, which provides a useful measure of the propensity of a α-helical peptide to interact with membranes; this is relevant, for example, to antimicrobial activity. The hydrophobic moment for a peptide with N residues is defined as

$$\mu_H = \left\{ \left[\sum_{n=1}^{N} H_n \sin(\delta n) \right]^2 + \left[\sum_{n=1}^{N} H_n \cos(\delta n) \right]^2 \right\}^{1/2} \qquad (1.5)$$

Here H_n is the hydrophobicity index (chosen from a suitable scale such as the Kyte–Doolittle or Wimley–White scales, see Table 1.1) for residue n, and δ is the angle of rotation of side chains around the backbone ($\delta = 100°$ for an α-helix).

Table 1.1 shows that some amino acids have similar properties. This can be further quantified in terms of physicochemical parameters used to assess amino acid replacement similarity. These are typically used to assess protein substitutions. Scales include the Grantham distance, Sneath's index, and others. These scales are based on parameterization of amino acid characteristics such as molar volume and polarities.

The tendency for residues to form α-helices or β-sheets has also been quantified, based on analysis of protein databases. A variety of scales have been proposed, which can be used in the design of peptides targeting these secondary structures. Figure 1.9 shows an example of one scale. Other scales

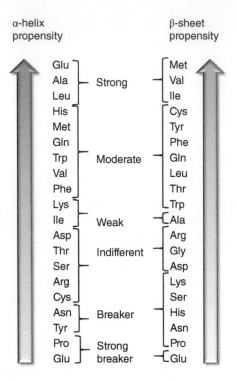

Figure 1.9 Chou–Fasman scale of α-helix and β-sheet propensity for protein and peptide residues.

to assess aggregation propensity for peptide sequences are available on the web and are listed in Table 1.6.

Amino acids that bear functional groups which can be removed or functionalized are substrates for enzymes. Residue-specific enzymes have names based on the amino acid appended with '-ase', for example arginase, asparaginase, and tyrosinase. Another class of residue-specific enzymes are kinases and phosphatases that respectively add or remove phosphoryl groups to tyrosine, serine, or threonine. Transferases have activity, as the name suggests, in transferring functional groups such as methyl groups, on to specific residues in proteins and peptides. Other examples of residue-specific enzymes include amino acid hydrolases and dehydroxylases. Enzymes such as proteases recognize specific configurations of amino acids in peptide/protein substrates. An important class, for example, are the serine proteases, which contain catalytic triads of residues for example Ser-His-Asp (these are not sequential but the notation means that these residues are present in specific configurations at the catalytic site) in chymotrypsin. Other reactions that exploit the functional groups on charged, hydrophilic, and tyrosine residues are discussed in Section 2.11.

Table 1.3 lists some non-standard (mostly non-proteinogenic) amino acids found in natural products and mentioned elsewhere in the book, among others. An extra $-CH_2-$ group between the carbonyl and amine groups in the backbone is characteristic of β-amino acids such as β-alanine. Two $-CH_2-$ groups separate the carbonyl and amine groups in γ-amino acids such as γ-aminobutyric acid, GABA (which is an important neurotransmitter) or statine (present in pepstatin, see Section 5.2). δ-Amino acids have three added $-CH_2-$ groups. Aminohexanoic acid (also known as aminocaproic acid) is an analogue of lysine with five $-CH_2-$ groups in the backbone instead of the side chain. Aminohexanoic acid is an important intermediate in the synthesis of nylon-6. The homo-amino acids are analogues of amino acids with additional $-CH_2-$ groups in the side chain.

Other residues in Table 1.3 are designed specifically to reduce the susceptibility of the peptide in which they are incorporated to proteolysis. Examples, other than the β-, γ-, and δ-amino acids, include the dehydro amino acids and amino acids containing unsaturated rings, as well as amino acids with blocked functional groups such as N-ε-methyllysine, N-methylglycine (sarcosine), which is strictly a peptoid residue (see Section 2.7).

Phosphorylation and dephosphorylation of tyrosine and serine are important in many biochemical reactions and these processes are achieved via enzymes. Kinases perform phosphorylation and are important targets for cancer therapies since kinases are involved in cell signalling, metabolism, and protein regulation. Activity of the enzyme tyrosine hydroxylase on tyrosine produces L-DOPA (L-3,4-dihydroxyphenylalanine) (see Table 1.3), which is used as a treatment for Parkinson's disease. This amino acid is an important precursor of the catecholamine neurotransmitters including dopamine (which results from removal of the carboxyl group from L-DOPA), norepinephrine (noradrenaline), and epinephrine (adrenaline). The enzyme tyrosinase acting on tyrosine also produces L-DOPA as an intermediate oxidant on the pathway to dopaquinone and its polymerization to produce melanin (Section 3.7). L-Ornithine (Table 1.3) is an analogue of lysine, having one extra methyl group in the side chain. It is produced by the activity of arginase on L-arginine.

There are only a few amino acids that contain elements other than the standard C, H, N, O, and S (or P in the case of the phosphorylated amino acids). Selenocysteine is the analogue of cysteine in which selenium replaces sulfur. It incorporates selenium into proteins *in vivo* and is considered to be the 21st proteinogenic amino acid.

Table 1.3 Some non-standard amino acids.

Symbol	Name	Structure
Abu	1-aminobutyric acid	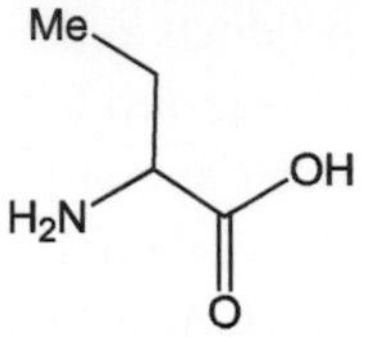
Ahx	6-aminohexanoic acid (6-aminocaproic acid)	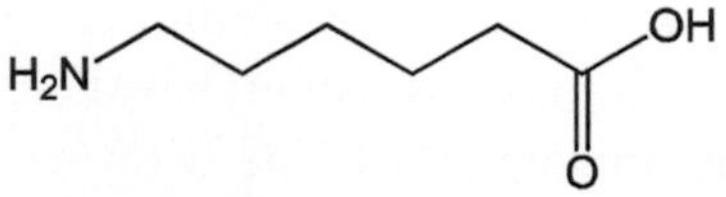
Aib	α-aminoisobutyric acid	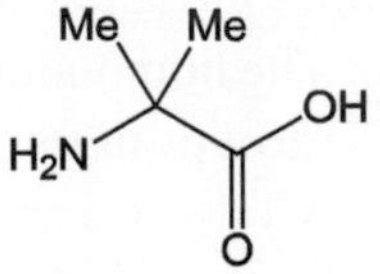
β-Ala	β-alanine	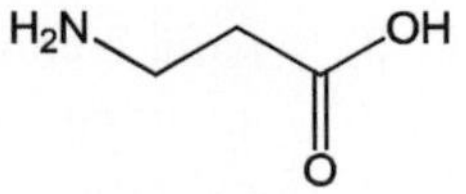
β-Asp	β-aspartic acid	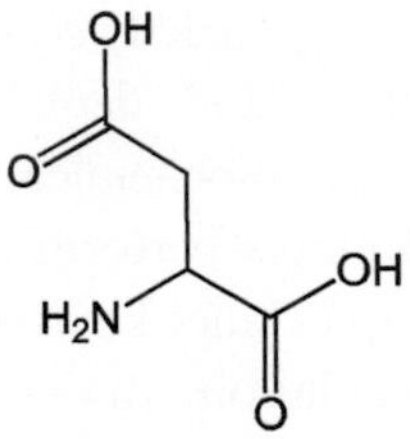
AzGly	aza-glycine	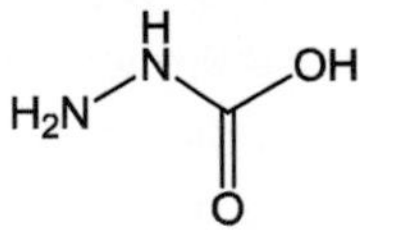
2-Bmt	(*E*)-2-butenyl-4-methylthreonine	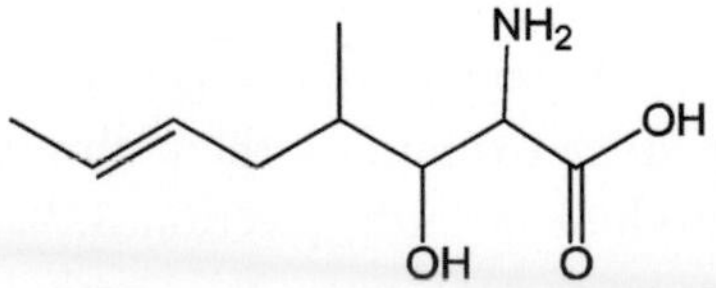
Cha	β-cyclohexylalanine	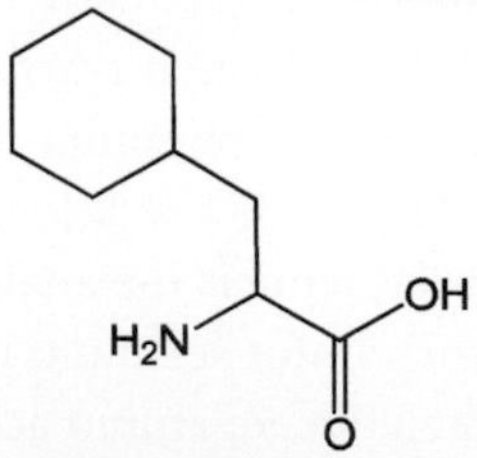

Table 1.3   (*continued*)

Symbol	Name	Structure
Chg	α-cyclohexylglycine	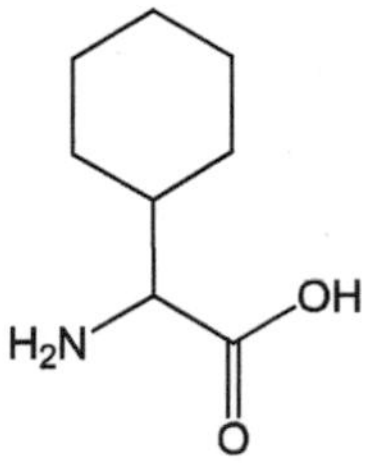
Cit	citrulline	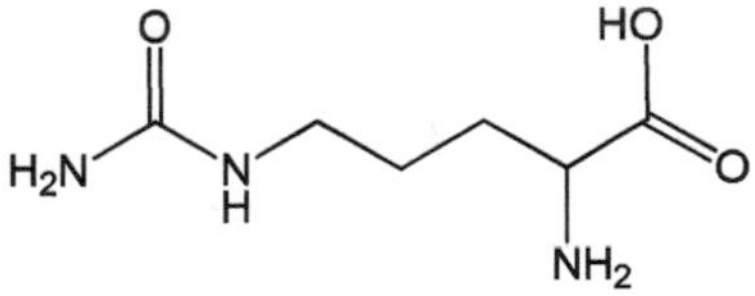
Dha	dehydroalanine	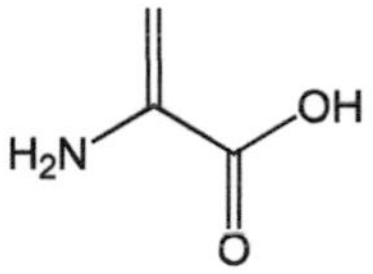
Dab	diaminobutyric acid	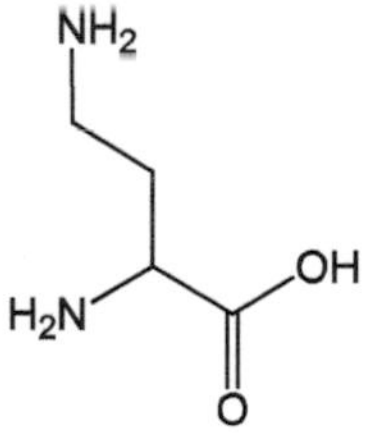
Dap	diaminopimelic acid	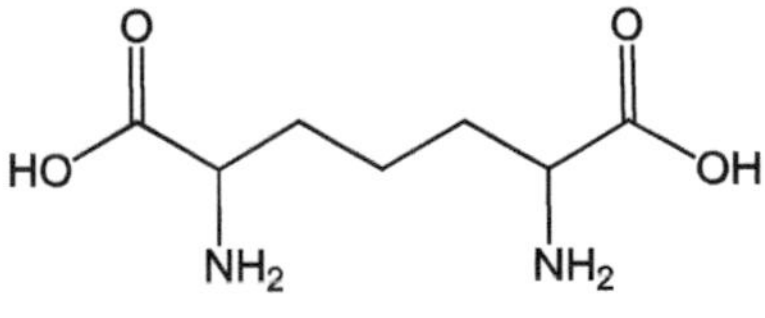
Dhb	2,3-dehydroaminobutyric acid	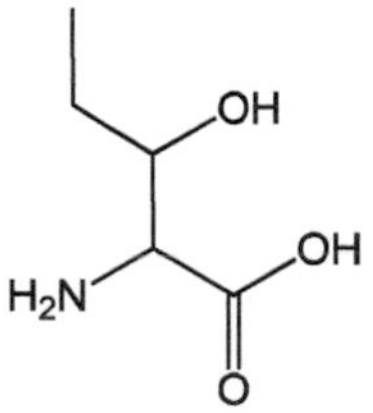

(*continued*)

Table 1.3 (*continued*)

Symbol	Name	Structure
DOPA	3,4-dihydroxyphenylalanine	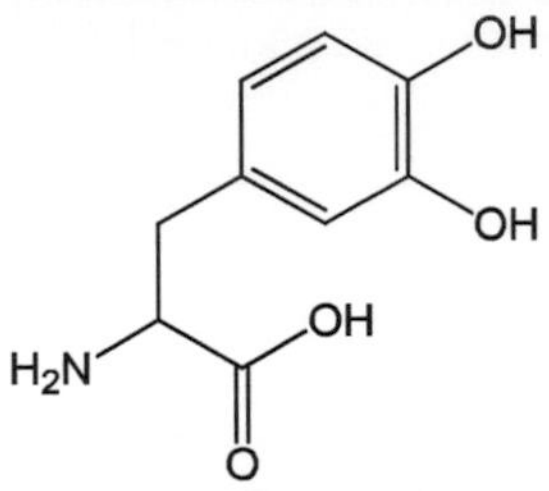
GABA	γ-aminobutyric acid	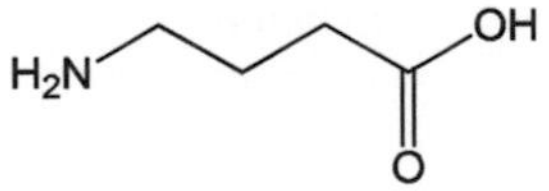
Gla	γ-carboxyglutamic acid	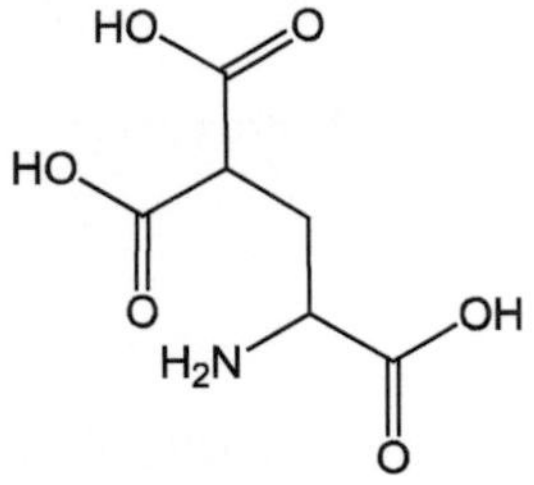
HArg	homoarginine	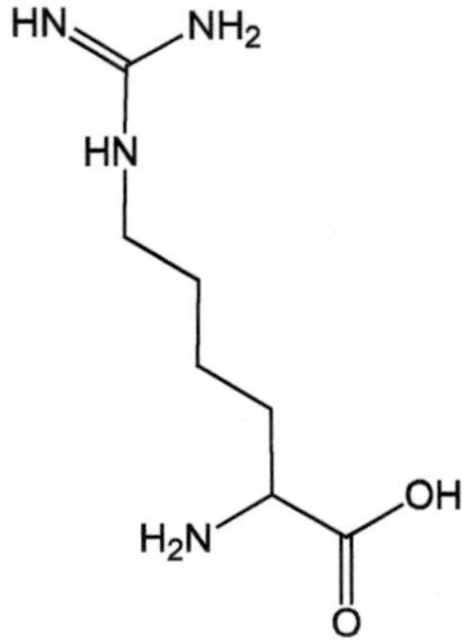
HCy	homocysteine	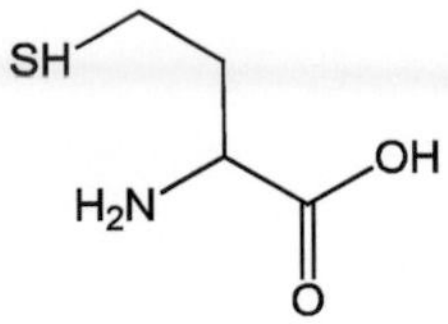
HSe	homoserine	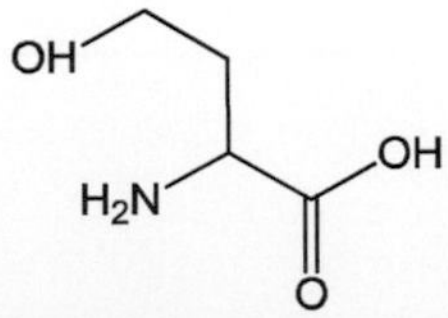

Table 1.3 (*continued*)

Symbol	Name	Structure
Hylys	hydroxylysine	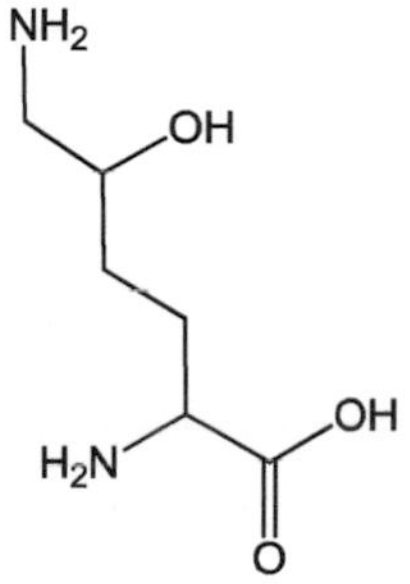
Hyp	hydroxyproline	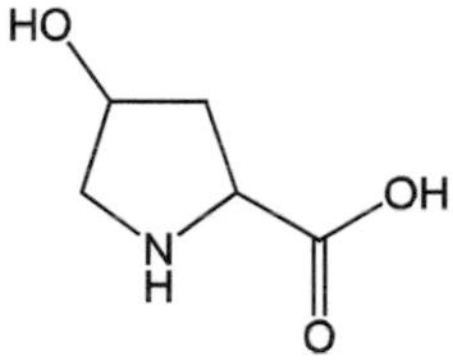
IsoGlu	isoglutamic acid (amino acid isomer)	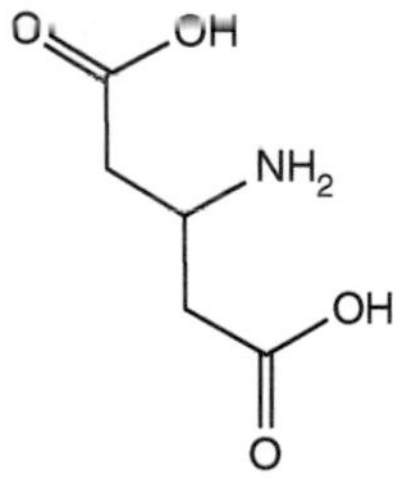
Kyn	kynurenine	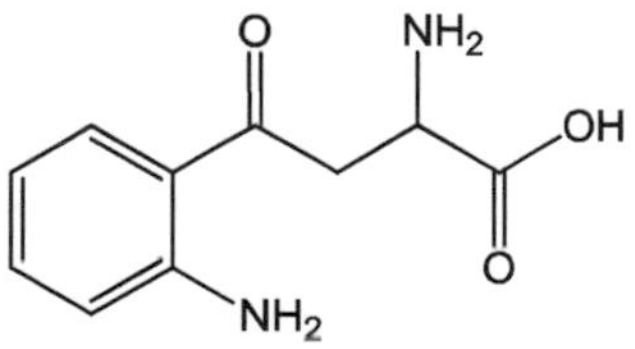
Lan	lanthionine	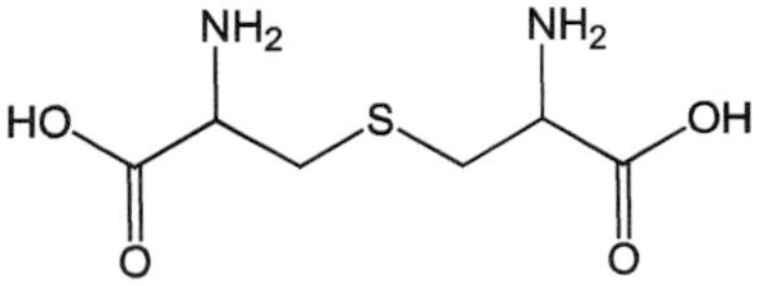

(*continued*)

Table 1.3 (*continued*)

Symbol	Name	Structure
Lys(Me)	N-ε-methyllysine	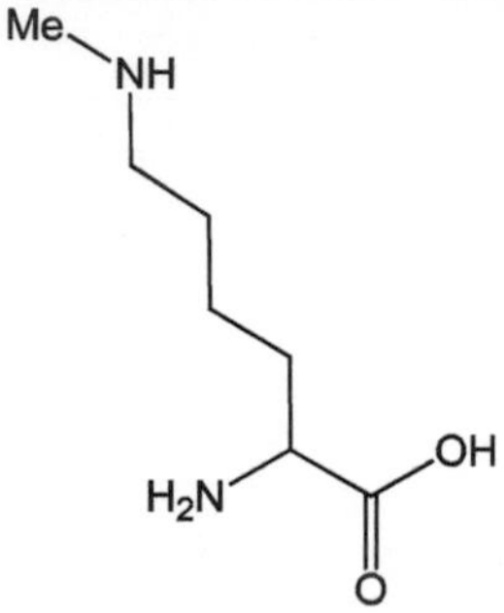
1-Nal	1-naphthylalanine	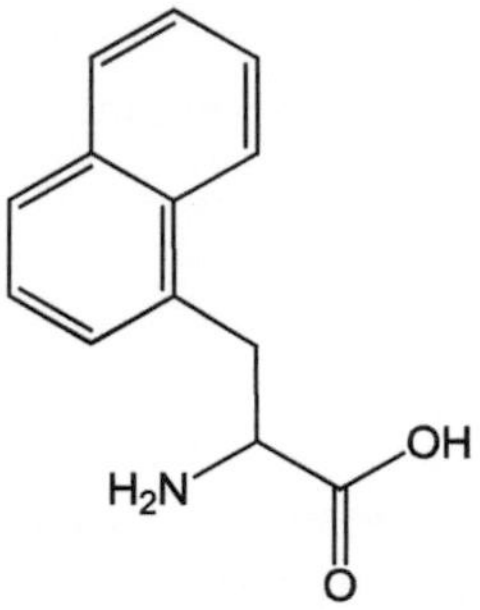
2-Nal	2-naphthylalanine	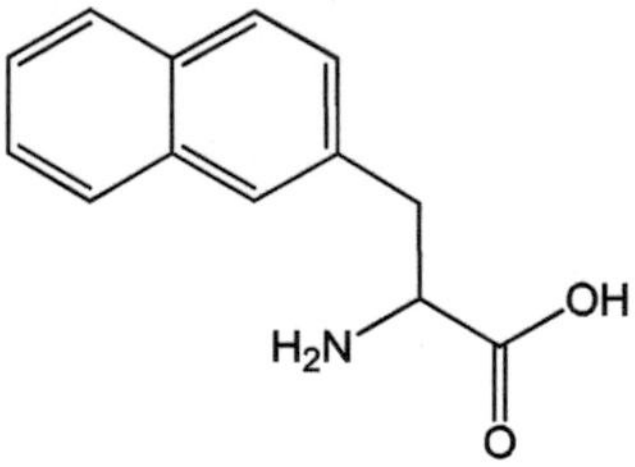
Nle	norleucine	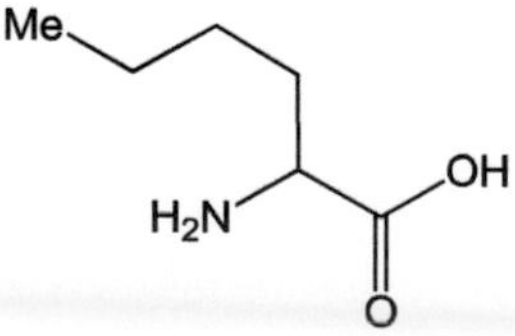
Nva	norvaline	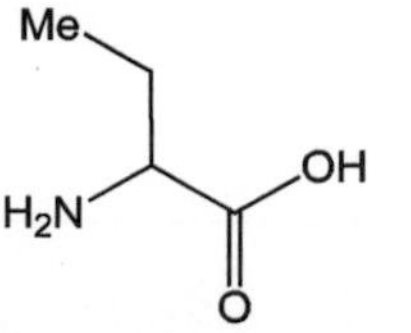

Table 1.3 (*continued*)

Symbol	Name	Structure
Orn	ornithine	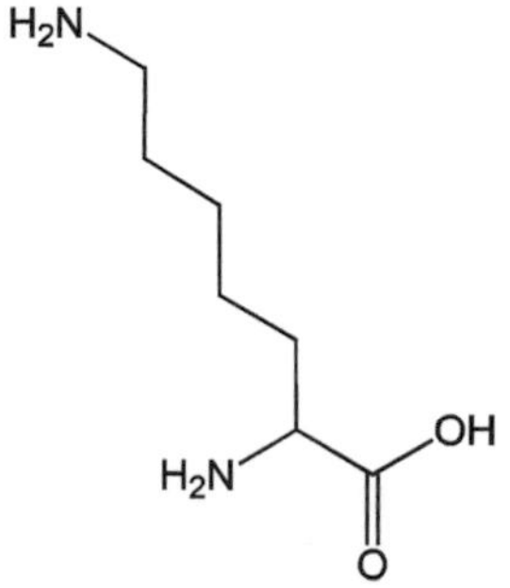
2-Pal	3-(2-pyridyl)alanine	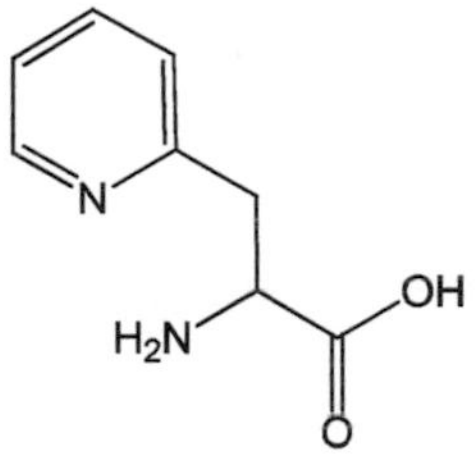
3-Pal	3-(3-pyridyl)alanine	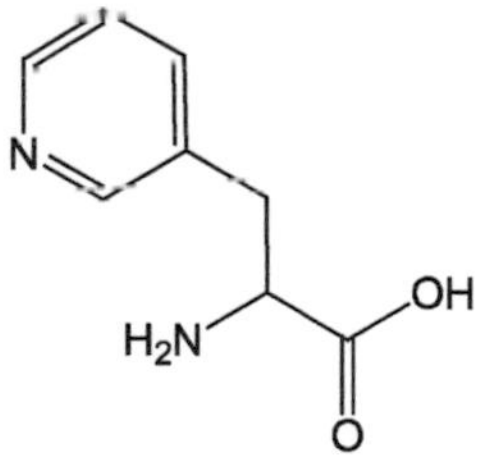
Pen	penicillamine	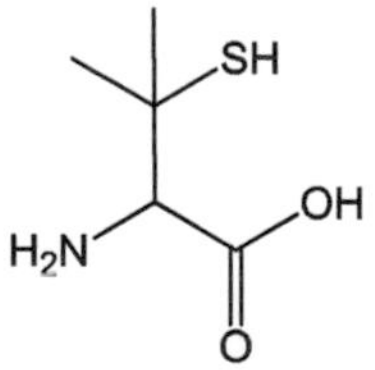
4-Cl-Phe	4-chlorophenylalanine	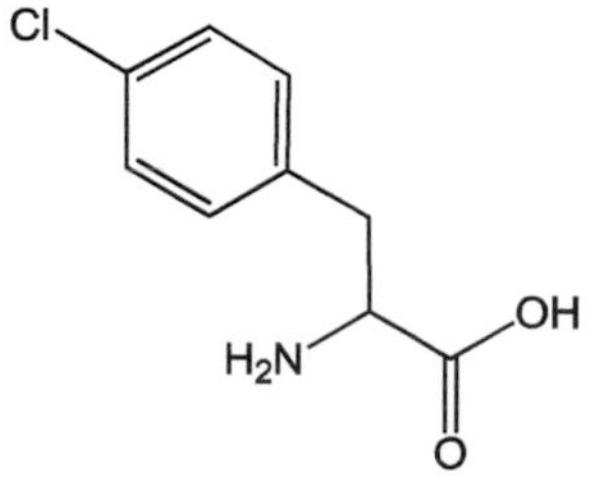

(*continued*)

Table 1.3 *(continued)*

Symbol	Name	Structure
4-NO$_2$-Phe	4-nitrophenylalanine	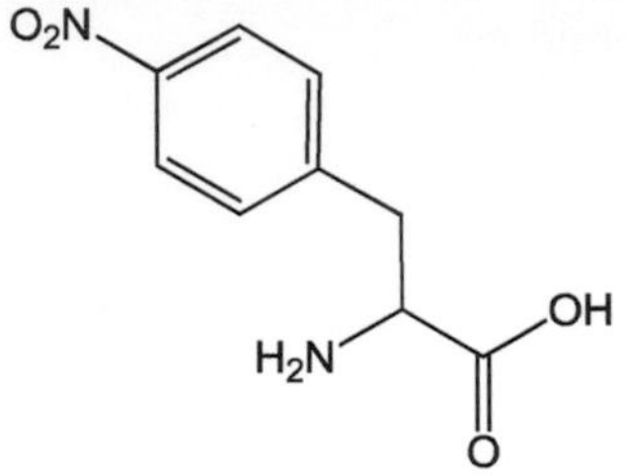
Phg	phenylglycine	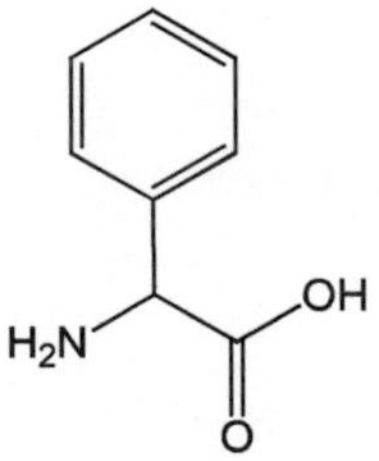
pGlu or Pyr	pyroglutamic acid	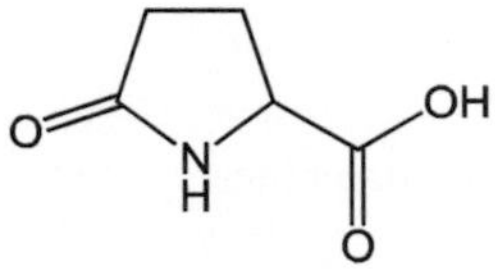
Pyl	pyrrolysine	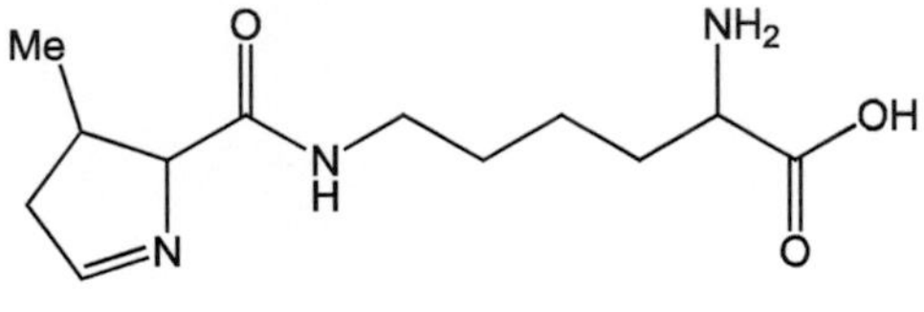
Sar	sarcosine	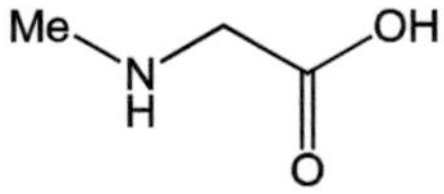
Sec	selenocysteine	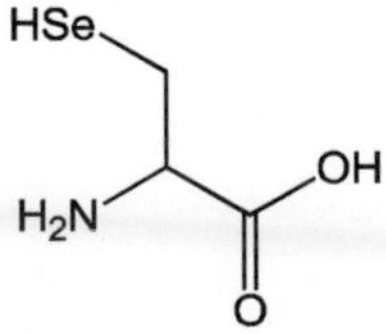
Sta	statine	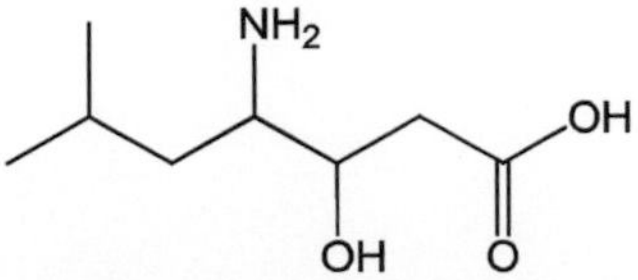

(continued)

Table 1.3 (*continued*)

Symbol	Name	Structure
4-ClThr	4-chlorothreonine	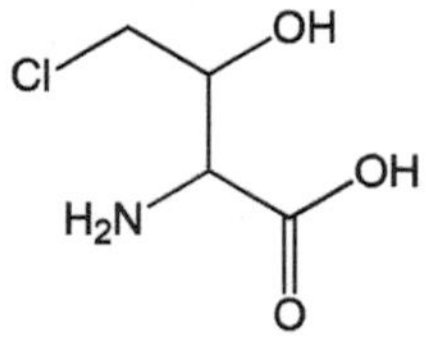
D-2-Me-Trp	D-2-methyltryptophan	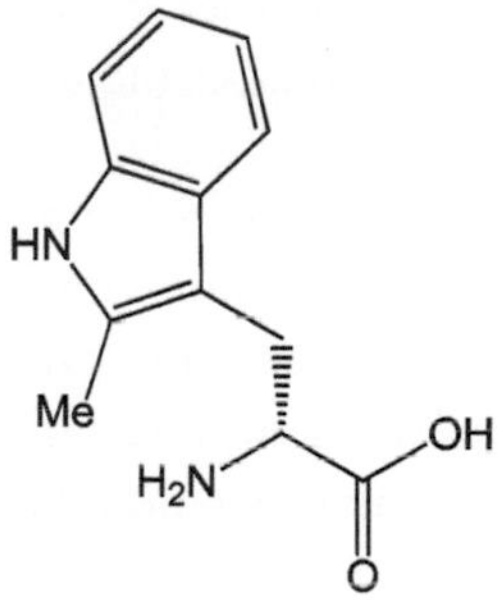
pSer	phosphoserine	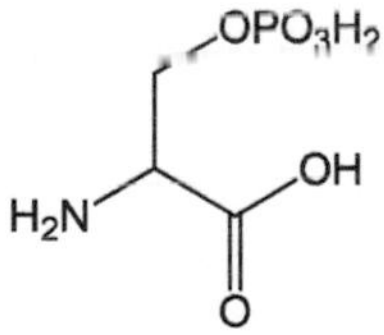
pThr	phosphothreonine	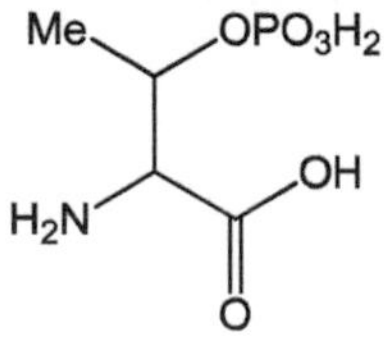
pTyr	phosphotyrosine	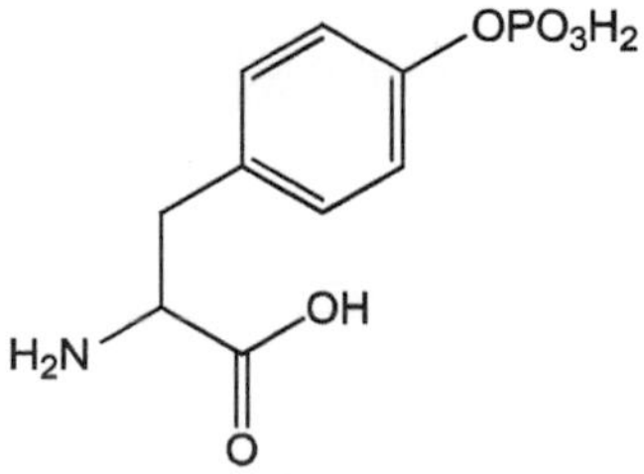

Table 1.3 includes one example (dehydroalanine) of a dehydroamino acid (one which contains a double bond in the side chain) although others have been used as non-natural residues resistant to proteolysis. Dehydroamino acids also occur naturally, being (rare) post-translational modifications. Analogues of aromatic amino acids with bulkier side chains, such as naphthylalanine, are included in Table 1.3, as are saturated variants such as α-cyclohexylalanine, an analogue of phenylalanine. Phenylglycine is a variant of Phe which lacks a spacer in the side chain. (There are no natural aromatic residues without a methyl spacer before the aromatic group.) Other entries in Table 1.3 include analogues of aromatic residues with substituents on the aromatic group, e.g. chloro- or nitro-substituted phenylalanine, or variants with other aromatic moieties, such as pyridyl groups or methyl-substituted tryptophan.

Norleucine and norvaline, shown in Table 1.3, are isomers of the natural amino acids leucine and valine respectively, lacking the side group branches present in the natural residues.

Among the other non-natural residues in Table 1.3, lanthionine is the product of the addition of cysteine to dehydroalanine. It is a key residue in the lantibiotics discussed in Section 4.6.1. Other amino acids in Table 1.3 are encountered rarely, and are included for completeness; in particular they are present in specific peptide therapeutic molecules mentioned in Chapter 5, where their contextual usage is explained.

1.3   THE PEPTIDE BOND

Nature produces peptides by linking amino acids; this takes place at the ribosome. Amino acids are coded by triplets of nucleic acids (codons), and considering that mRNA (messenger RNA) is translated, triplets of RNA bases specify particular amino acids. This is the standard genetic code. Of course, peptides can now easily be synthesized in the laboratory using methods discussed in Chapter 2.

The peptide bond is formed in a condensation (dehydration) reaction between the carboxyl group of one amino acid and the amine group of another (Figure 1.10). The residues in a peptide are linked through amide units, which are planar.

The geometry around a peptide bond is shown in Figure 1.11. An important fact about peptide bonds is the planarity of the amide C=O–NH unit, which constrains the possible conformations that the chains can adopt. On average, The C–N bond length is 1.33 Å, the N–C_α bond length is 1.45 Å,

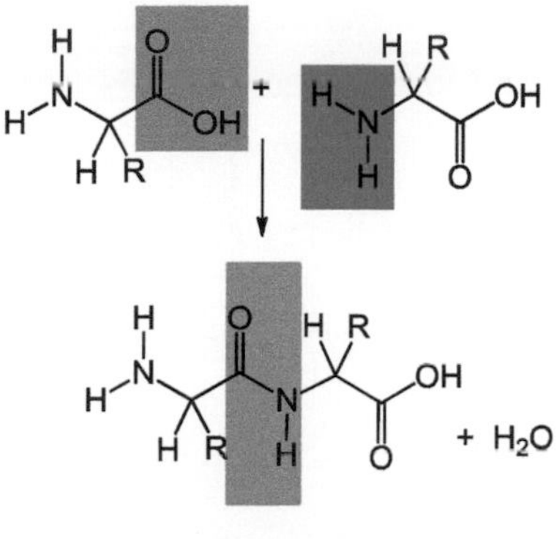

Figure 1.10 Formation of a peptide bond (amide group highlighted in purple) from condensation reaction between the carboxyl group (red) of one amino acid and the amino group (blue) of another.

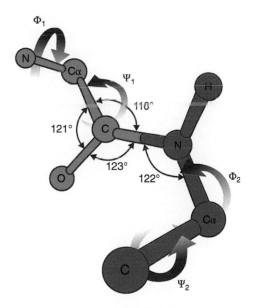

Figure 1.11 Peptide bond geometry with typical angles shown.

and C–C$_\alpha$ is 1.52 Å. The angles Φ are defined by rotation about the N–C$_\alpha$ bonds while Ψ is defined by rotation around C$_\alpha$–C. The case Φ = 0 defines the case where the C$_\alpha$–C bond is *trans* to the N–H bond, whereas Ψ = 0 corresponds to C$_\alpha$–N *trans* to C=O. Not shown in Figure 1.11 for simplicity is the angle Ω, the rotation angle about the N–CO bond; this angle is usually 180°.

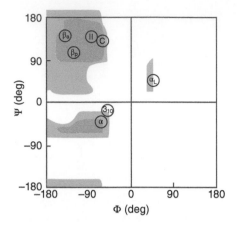

Figure 1.12 A Ramachandran plot. The shaded areas show the 'normally allowed conformations' in pink and the 'outer limit' structures in blue. Symbols as follows: β_a is antiparallel β-sheet, β_p is parallel β-sheet, II is polyproline II/polyglycine II, C is collagen (a triple helical), 3_{10} is a 3_{10}-helix, α is a (normal right-handed) α-helix, and α_L is a left-handed α-helix.

The peptide bond links amino acids in polypeptide chains, which nature has evolved to exploit as scaffolds that present sequences of residues with the right degree of flexibility to adopt configurations corresponding to different secondary structures that provide the structure and function of proteins and peptides.

Peptide secondary structures (discussed further in Section 1.4) are characterized by particular favoured torsional angles, as summarized in a Ramachandran plot of conformational space (Figure 1.12). Such a plot is also widely used for protein conformational analysis.

Conventionally (as in this book), peptide sequences are written or drawn from the N-terminus on the left to the C-terminus on the right.

1.4   SECONDARY STRUCTURES

The most common ordered peptide secondary structures are the α-helix and β-sheet. These are also common elements of protein tertiary structure. Figure 1.13 shows an α-helix and the two types of β-sheet structure, with β-strands (peptide chains within β-sheet structures) aligned parallel or antiparallel to one another. The α-helix is stabilized by intramolecular hydrogen bonding, in contrast to the intermolecular hydrogen bonding of β-sheets. The standard α-helix is a right-handed structure, although the

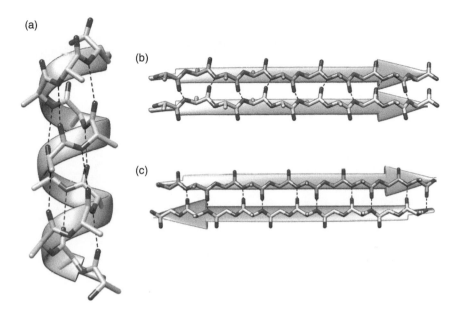

Figure 1.13 Principle regular peptide secondary structures: (a) α-helix (shown as a chain representation superposed on a ribbon representation) and (b, c) β-sheet, including (b) parallel and (c) antiparallel β-sheet structures, superposed on ribbon representations. Hydrogen bonds are shown as dashed lines.

non-standard left-handed α_L structure is also known. Figure 1.12 shows the corresponding region of the Ramachandran diagram. The α-helix may also be represented as a projection along the axis, in a so-called helical wheel diagram. An example is shown in Figure 1.14. Due to the patterning of the carbonyl groups, the α-helix structure has a net dipole moment. Typical structural parameters for regular peptide conformations are listed in Table 1.4. As an example, a decameric (10-mer) peptide adopting an antiparallel β-sheet structure would be calculated to be $10 \times 3.4 = 34$ Å long, while the same peptide forming a parallel β-sheet would be slightly shorter (32 Å). In a fully extended peptide chain, $\Phi = \Psi = \Omega = 180°$.

As well as the ordered secondary structures, peptides can also adopt a disordered structure or coil structure, which is less rigorously known as a random coil structure. The term 'random coil structure' is not recommended since peptides will not adopt a truly random coil due to the finite volume of the chain and steric interactions between side chains.

The transition from α-helix to disordered (or vice versa) can be measured as a function of temperature (or chain length). The transition typically shows a sigmoidal-shaped profile (Figure 1.15) which can be described using the Zimm–Bragg model. This allows for the cooperativity of the transition

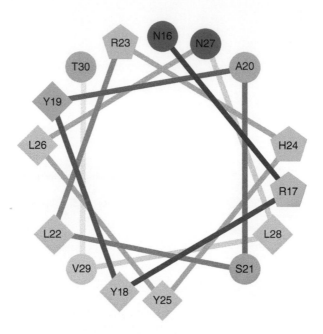

Figure 1.14 Example of a helical wheel diagram of a α-helical peptide sequence. This is a representation of the α-helical domain of peptide PYY_{3-36} (discussed further in Section 5.3.5). The hydrophilic residues are represented as circles, hydrophobic residues as diamonds, and positively charged residues as pentagons. Hydrophobicity is colour coded in shades of green. The positively charged residues are blue. Hydrophilic uncharged residues are coded red-yellow, with pure red being the most hydrophilic residue, and the amount of red decreasing proportionally to the hydrophilicity. This figure was generated using one of several web servers able to generate these diagrams.

Table 1.4 Bond angles and residue spacings for regular peptide conformations.

Structure	Bond angle (deg)			Residues per turn	Translation per residue (Å)
	Φ	Ψ	Ω		
Antiparallel β-sheet	−139	135	−178		3.4
Parallel β-sheet	−119	113	180		3.2
Right-handed α-helix	−57	−47	180	3.6	1.50
3_{10}-helix	−49	−26	180	3.0	2.00
π-helix	−57	−70	180	4.4	1.15
Polyproline II	−78	149	180	3.0	3.12
Polyglycine II	−80	150	180	3	3.1

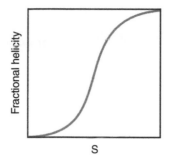

Figure 1.15 The coil–helix transition is cooperative, leading to a sigmoidal shaped profile.

between the two states at the level of each residue which is described by the probability $\sigma \times s$ that a disordered residue (rather confusingly termed a 'coil residue', referring to random coil) is followed by a helical residue (or vice versa). Here $s = [H]/[C]$ is an equilibrium constant expressed in terms of helix and coil concentrations, and σ is a nucleation probability (for the coil state). This leads to a 2×2 matrix of conditional probabilities describing the probabilities for pairs of residues. The model (which is a one-dimensional lattice model) can be computed analytically. The Lifson–Roig model is related to the Zimm–Bragg model but allows for the fact that in α-helices the hydrogen bonds occur after every three residues. The conditional probabilities for residue pairs are represented in a 4×4 transfer matrix. These models can be extended to describe transitions to other helical states (described below), other than α-helical. The Zimm-Bragg model can quantify the average degree of helicity but shows that this can comprise a significant fraction of partially helical chains even if (as in the case of short peptides) the fraction of fully helical sequences is very small.

The β-sheet structure is said to be 'pleated' since C_α atoms are successively above and below the plane of the β-sheet. Most β-sheet structures in proteins and many β-sheet amyloid structures of peptides are twisted. The twist is usually right-handed and arises from relative rotation of residues (the degree of rotation depends on the side chains). Parallel β-sheets are less twisted than antiparallel ones. Antiparallel β-sheet structures are considered more stable for this reason and due to the more regular hydrogen bond geometry (see Figure 1.13).

A strategy to design peptides that adopt β-sheet structures uses alternating residues, especially alternating aliphatic residues and charged residues, e.g. AKAKAK or ELELEL. In some cases, for example silk (discussed further in Section 3.7), alternating residues without charges, e.g. GAGAGA, can favour β-sheet structures.

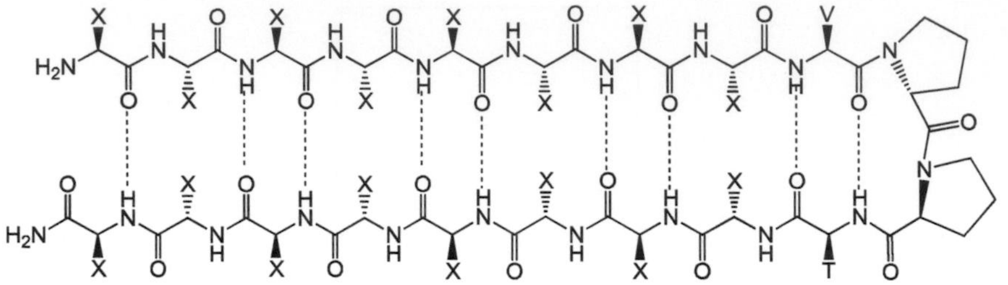

Figure 1.16 Example of a β-hairpin structure.

Where antiparallel β-sheets are required by design, then it is possible to use a β-hairpin motif. An example is that shown in Figure 1.16, which contains intramolecular hydrogen bonds as well as intermolecular hydrogen bonds. β-hairpins are structures with short 2–5 residue loops. A common motif employs the V^DPPT motif, as in Figure 1.16. Other methods use non-peptide linkers to control the alignment of strands within molecules to provide a constrained intramolecular antiparallel structure. Proline and glycine are common residues in β-hairpin structures (and also turn structures) since they allow the chain to take up this unusual conformation.

In addition to the α-helix and β-sheet, a number of other regular structures may be adopted by peptides, including helical structures. Examples are shown in Figure 1.17, such as polyproline II (PPII), polyglycine II (PGII), 3_{10} and π helices, and collagen (triple helical) structures. Despite its name, the helical PPII structure can be observed for peptides that do not have a polyproline backbone.

A variety of turn structures are known for peptides and are common at or near the surface of proteins where the peptide chain needs to fold. Turn structures are those in which the backbone changes direction, nearly reversing direction in the case of the main types of turn structures, β- and γ-turns. Many types of turn structure have been identified, including α-, δ-, and π-turns as well as at least nine types of β-turns and two types of γ-turns. These have different backbone torsion angles. Figure 1.18 shows some examples of β- and γ-turn structures; β-turn structures have two non-hydrogen bonded residues in the turn, whereas γ-turns have only one non-hydrogen bonded residue.

The unusual α-sheet structure has recently been reported for several peptides. The structure is related to that of the β-sheet, but all carbonyl groups in a strand are oriented on one side of the sheet, while all the amino groups are oriented on the other side.

Peptide secondary structures are stabilized primarily by hydrogen bonds. Other intermolecular interactions play a role in specific structures and in

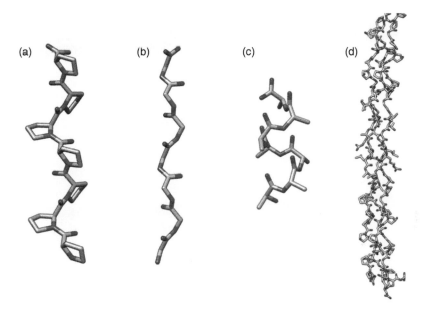

Figure 1.17 Helical peptide secondary structures. (a) PPII structure, (b) PGII structure, (c) 3_{10}-helix (for illustrative purposes only, shown with alanine residues), (d) collagen triple helix.

stabilizing protein tertiary structures. Figure 1.19 shows key interactions that stabilize secondary and tertiary structures. Disulfide bonds are present in many proteins, providing thermal conformational stability. Other amino acids, including tyrosine, can also form cross-links; two tyrosine residues can form dityrosine at high pH. Disulfide bridges are formed oxidatively and be broken by using reducing agents such as DTT (dithiothreitol). Salt bridges stabilize folded protein conformations and coiled coil structures. Hydrophobic interactions stabilize folded tertiary structures and are important in β-sheet aggregation. Aromatic π-stacking interactions are also important in the stabilization of β-sheet amyloid structures, discussed in Chapter 3.

Coiled coil peptides are aggregates of α-helices with facial amphiphilicity. The peptide sequence in each coil in many designed coiled coil peptides may be represented as a heptad abcdefg. For example, the peptide represented by the helical wheel diagram in Figure 1.20a has a hydrophobic face and a partly charged face, and these helices are likely to aggregate into dimeric structures. The packing of α-helices into dimeric coiled coils can also be driven by purely side chain interactions, as in the 'knobs-in-holes' model, proposed by Crick and shown in Figure 1.20b. This model is also

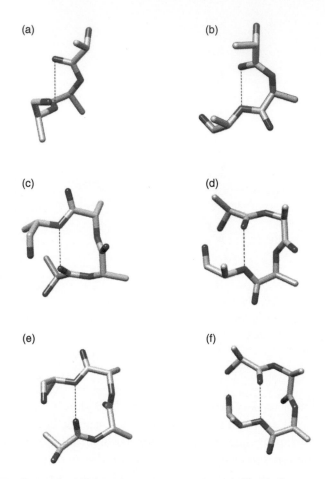

Figure 1.18 Examples of β- and γ-turn structures. (a) Classical γ-turn, (b) inverse γ-turn, (c) type I β-turn, (d) type I′ β-turn, (e) type II β-turn, (f) type II′ β-turn. The dashed lines indicate hydrogen bonds.

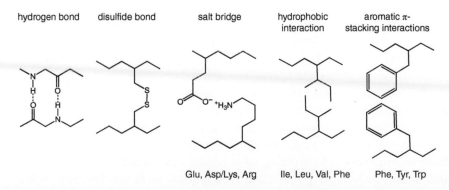

Figure 1.19 Backbone and side chain interactions that stabilize peptide structures.

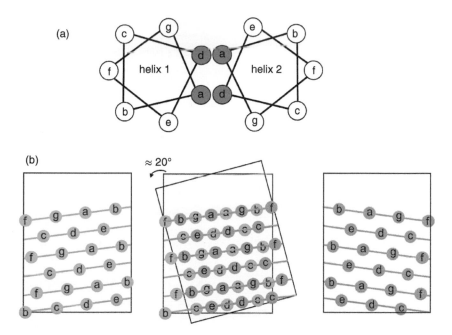

Figure 1.20 (a) A dimeric coiled coil structure based on the abcdefg heptad in α-helical peptides is favoured, for example, by E and K residues in positions d and a (green circles) while the other residues may be hydrophobic. (b) 'Knobs in holes' packing of side chains in a coiled coil dimer.

important in interpreting helix bundle structures in proteins. Figure 1.20b shows projections of the side chains of two helical peptides onto a plane parallel to the helix axes – a so-called helical net diagram. Crossing of the side chains at an angle of about 20° leads to efficient packing of 'knobs in holes'.

A special class of coiled coil stabilized by regular leucine side chain packing are the leucine zipper structures, which are important structural motifs in proteins. Every seventh residue is leucine. Figure 1.21 shows a leucine zipper structure called bZip (basic-region leucine Zipper) which templates (anionic) DNA during transcription in eukaryotes through cationic lysine and arginine residues in the N-terminal region.

Higher-order coiled coil structures are possible (helix bundles) and the sequence can of course be designed to favour such structure by substitution of residues in suitable positions, to favour salt bridging (E–K interactions) and/or hydrophobic interactions. The engineering of coiled coil peptide structures is a fascinating topic that is the subject of ongoing research.

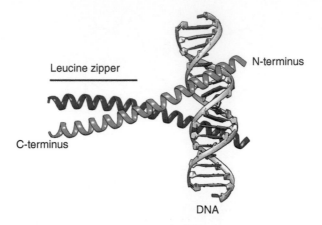

Figure 1.21   Leucine zipper (bZip) structure involved in DNA transcription.

1.5   PEPTIDE STRUCTURE AND CONFORMATION CHARACTERIZATION METHODS

1.5.1   UV/vis Absorbance

For peptides containing tyrosine and/or tryptophan and/or cysteine, the concentration can be obtained from UV absorbance measurements based on the known molar absorptivities (extinction coefficients, ε) of the three aromatic residues. Measurement of UV absorbance at 280 nm enables the extinction coefficient calculated from

$$\varepsilon = (m\text{W} \times 5500) + (n\text{Y} \times 1490) + (p\text{C} \times 125) \tag{1.6}$$

where m, n, and p are the numbers of tryptophan, tyrosine, and cysteine residues, respectively, and 5500, 1490, and 125 are residue absorptivities in units of M^{-1} cm^{-1}. Concentration can then be obtained via the Beer–Lambert law.

For peptides lacking Y or W residues, a common method to determine concentration is to measure UV absorbance at 205 nm, and use an average $\varepsilon(205 \text{ nm}) = 31$ ml mg^{-1} cm^{-1} (for example). This method (or using the above formula at 280 nm) is used in commercial UV concentration measurement instruments.

1.5.2   Circular Dichroism Spectroscopy (CD)

Circular dichroism (CD) is a technique to probe the secondary structure of chiral biomolecules, including peptides. The method relies on the differential absorption of right and left circularly polarized light. The usual method is based on 'fingerprinting' of spectral features in the 190–250 nm far-UV region. Data in the near-UV region can provide information on the conformation of peptides containing aromatic residues with absorption features in the 250–310 nm range. In the far-UV region, characteristic minima are observed in the absorption spectra at (approximately) 208 nm and 222 nm (α-helix) or 216–220 nm (β-sheet), as shown in Figure 1.18. On the other hand, a broad weak minimum in the range 195–200 nm is characteristic of disordered, sometimes known as random coil, conformation. This can be contrasted with the spectra for the polyproline II (PPII) conformation, which is characterized by a minimum in the CD spectrum at around 190–205 nm, along with a broad positive maximum at around 215–225 nm (Figure 1.22).

The CD spectra for the main secondary structures are characterized by typical values of the molar (or alternatively mean residue) ellipticity, as well as the position of the maxima/minima. In fact, CD spectra should be normalized (Eq. (1.7)) for this reason, and also to facilitate comparison between

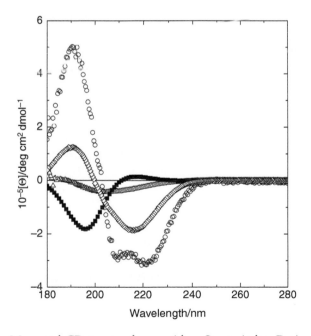

Figure 1.22 Measured CD spectra for peptides. Open circles: Derivative of peptide PYY_{3-36} (discussed in detail in Section 5.3.5) α-helix. Source: Taken from V. Castelletto et al. 2018, diamonds: β-sheet spectrum of lipopeptide C_{16}-KTTβAH Source: Taken from V. Castelletto et al. 2019, closed squares: PPII structure of RAAAAAAAAAR Source: Taken from C. J. C. Edwards-Gayle et al. 2019, open triangles: disordered structure of IKPEAP Source: Taken from J. A. Hutchinson et al. 2019.

samples at different concentrations or measurements in different path length cells. The molar ellipticity is given by

$$[\theta] = \frac{\theta}{10cl} \tag{1.7}$$

Here $[\theta]$ is the molar ellipticity (in units deg cm^2 dmol^{-1}), θ is the measured CD signal amplitude in millidegrees (mdeg), c is the molar concentration, and l is the path-length of the cell in cm. The magnitude of the mean residue ellipticity (MRE, molar ellipticity divided by number of residues) at 222 nm can be used to determine the α-helical content, f_α, of coil peptides via the equation:

$$f_\alpha = 100\,[\theta]_{222}/[\theta]_{222}^{ex} \tag{1.8}$$

$[\theta]_{222}^{ex}$ in Eq. (1.3) is the extrapolated value for the molar ellipticity:

$$[\theta]_{222}^{ex} = [\theta]^{\infty}\left(1 - \frac{k}{n}\right) \tag{1.9}$$

where $[\theta]_{222}^{ex} = -31\,556$ deg cm^2 dmol^{-1}, $[\theta]^{\infty} = -37\,400$ deg cm^2 dmol^{-1} is the maximum MRE at 222 nm of a peptide with infinite length and 100% helix content, n is the number of residues/helix, and k is a wavelength-dependent constant (2.5 at 222 nm). Comparison of the ratio of the molar ellipticity at 222 and 208 nm gives information on the coiled coil content. Values $[\theta]_{222}/[\theta]_{208} > 0.9$ typically indicate coiled coil formation.

CD spectra for proteins are usually analysed using algorithms based on databases compiled for proteins for which the X-ray crystal structure is known. A range of software is available based on various databases. Most consider only larger proteins although there are limited reference data sets (and curve-fitting programs) for shorter peptides. This type of analysis is of little use for peptides (especially short ones) for which individual residues (especially aromatic residues) and specific conformations (e.g. turns) can dominate the CD spectrum. There is as yet not a comprehensive theory to describe these effects. Theoretical interpretation of CD spectra involves complicated quantum mechanical analysis of electronic states, specifically of electric dipole and magnetic dipole transition moments.

Linear dichroism (LD) is a more specialized method, which is less commonly used than CD. LD refers to the differential absorbance of plane-polarized UV radiation, and it gives information on the alignment of extended objects resulting from peptide self-assembly. Since amyloid fibrils and nanotapes are highly anisotropic; they can align under flow or in other fields and this can be probed using LD, which in particular provides information on the orientation of the peptide backbone and of chromophores such as aromatic residues.

1.5.3 FTIR and Raman Spectroscopy

FTIR (Fourier transform infrared) spectra of peptides may be measured in transmission mode from solutions in a closed cell or from a drop deposited on a crystal in ATR (attenuated total reflectance) mode or from dried films (less preferred since the spectrum does not correspond to a peptide with conformation under the solution conditions). FTIR is sensitive to the vibrational modes of bonds within peptides. Specific regions of the spectrum are the focus of particular analysis.

The amide I region in the range $1600-1700\,\text{cm}^{-1}$ is sensitive to the modes of CO, CN, and NH groups, which are influenced by H-bonding. A prime is added to the name of the region of the spectrum, for example to give the term amide I' for spectra measured in D_2O. All band positions mentioned above are slightly downshifted if measurements are performed in D_2O. In fact, measurements in D_2O are beneficial since water absorption features around $1650\,\text{cm}^{-1}$ can be avoided. Characteristic FTIR bands of peptides and proteins are listed in Table 1.5.

The amide I region gives information on secondary structure, via 'fingerprinting' or peak-fitting methods. These are prone to uncertainty associated with the overlap of peaks in the spectra, although clear features of secondary structures such as β-sheets can be resolved by performing measurements in transmission mode on a modern FTIR instrument, with sufficiently concentrated samples in D_2O in narrow path-length cells. A peak typically in the $1620-1640\,\text{cm}^{-1}$ range is associated with β-sheet structures. A peak in the typical range $1648-1657\,\text{cm}^{-1}$ is associated with α-helix structure, whilst disordered peptides give a peak in the typical range $1642-1650\,\text{cm}^{-1}$. The

Table 1.5 Typical FTIR band ranges for peptides.

Band	Approximate wavenumber (range)	Assignment
Amide A	3270–3310	N–H stretch
Amide B	3030–3100	N–H stretch
Amide I	1600–1700	Mainly C=O stretch with contribution from C–N stretch and N–H bend
Amide II	1500–1600	Combination of N–H in-plane bend and C–N stretch with minor contributions from other modes
Amide III	1200–1400	N–H bend and C–H stretch, C=O stretch, OCN bend, and others

narrow intense band observed for some peptides at 1675–1695 cm^{-1} is usually ascribed to antiparallel β-sheet structure. Caution is required since a peak in the amide I region of an FTIR spectrum close to 1673 cm^{-1} is due to residual trifluoroacetic acid from the peptide synthesis, bound to peptide cations (unless this is removed by ion-exchange methods). Tables of FTIR peaks associated with side groups in some residues are available. A peak near 1705 cm^{-1} may be assigned to the carbonyl stretch, e.g. in acidic side residues or from the C terminus.

The amide II band around 1550 cm^{-1} mainly results from the N–H bending vibrations which are highly sensitive to deuteration (the deuterium from D_2O exchanges positions with hydrogen from the N–H bond). As a consequence the amide II′ band is shifted by approximately 100 cm^{-1} to 1450 cm^{-1} in D_2O.

Raman spectra provide similar information on peptide secondary structure and side chain deformation to FTIR, since the vibrational transition selection rules are generally satisfied for peptide deformation modes for both Raman and FTIR spectroscopy. As it is a scattering-based method, Raman spectroscopy also typically requires a higher sample concentration than FTIR to get a useful spectrum. However, measurement of Raman spectra can be recorded in H_2O, in contrast to FTIR where the water absorption band prevents measurements in the vital amide I region. Raman microscopy enables the chemical mapping of peptide samples scanning across the sample at micron resolution and measuring Raman spectra.

Isotope labelling of peptides enables specific features in the FTIR spectrum to be resolved, as bands associated with deformation modes of labelled residues will be shifted. Typically, isotopic substitution using a ^{13}C-labelled peptide splits amide I bands for β-sheets into higher and lower frequency peaks. The magnitude of the ^{12}C band wavelength shift depends on the extent of perturbation of the ^{12}C carbonyl coupling induced by ^{13}C substitution and strongly coupled β-sheets are generally sensitive to such isotopic substitution. This can be used to infer information on the registry of β-strands.

FTIR can be extended to study linear dichroism (polarized FTIR) on aligned samples with isotope labelling and to vibrational circular dichroism (VCD). VCD is analogous to (UV) electronic CD, but extended to the IR range of the spectrum, sensitive to bond vibrations. The analogous technique using Raman scattering is Raman optical activity (ROA). Polarized Raman spectroscopy can provide information on the orientation of specific features such as aromatic residues.

Recently, two-dimensional (2D) IR methods have attracted great interest as it is possible to correlate vibrational modes, providing information on mode coupling. When combined with isotope editing of peptides, 2D IR can provide detailed information on β-sheet conformation and the pathway of amyloid formation (a topic discussed further in Sections 3.2 and 3.4).

1.5.4 NMR Spectroscopy

This section does not consider the routine use of NMR (nuclear magnetic resonance) to analyse synthesized peptide structures, but is rather focused on specialized NMR techniques to probe peptide aggregates.

The secondary structure of a peptide or protein can be analysed via the chemical shift index (CSI), which compares the chemical shift of $^1H_\alpha$ or ^{13}C backbone units to those for the corresponding residue in an ideal disordered peptide. Figure 1.23 shows an example for the single immunoglobulin binding domain, which comprises 56 residues (including the NH_2-terminal Met) of protein G from group G *Streptococcus*. The CSI is consistent with the structure for this small protein which comprises four β-strands and an α-helix. Subsequent to the development of the CSI, other methods such as probability-based secondary structure identification (PSSI) have been introduced to analyse peptide chemical shift data.

Information on secondary structure can also be obtained from NOEs (nuclear Overhauser effects), since these depend on the separation between nuclei.

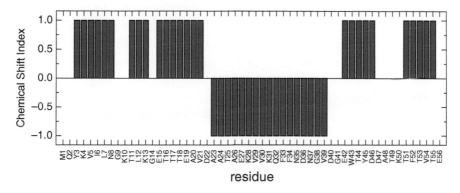

Figure 1.23   Chemical shift index plot (for $^1H\alpha$) for 56-residue sequence from protein G. Strands correspond to +1 states, helices to −1 and coil (disordered) to 0.

Two-dimensional NMR has become a standard method for more detailed analysis of peptides via correlation or exchange spectroscopic methods. The most commonly employed for peptide structure analysis are NOESY (nuclear Overhauser effect spectroscopy), TOCSY (total correlation spectroscopy), and DQF-COSY (double quantum filtered correlation spectroscopy) for homonuclear measurements and HSQC (heteronuclear single-quantum correlation spectroscopy) for heteronuclear studies (e.g. ^{1}H–^{15}N, ^{1}H–^{13}C). For peptides with fewer than 50 residues, peak assignments are possible using homonuclear 2D NMR methods; for longer peptides heteronuclear experiments using isotopically labelled samples can be used. These can also be used to probe strand alignment in β-sheet structures (composed of identical peptides) which are otherwise impossible to distinguish.

Amide proton exchange rates also provide information on secondary structure, since these are influenced by the hydrogen bonding that is characteristic of secondary structure formation.

Solution NMR can be used to provide high-resolution peptide structures, such as those of helical peptides, via 2D correlation methods – as described above. 2D correlation methods are used to provide distance and backbone torsional angle information that can then be used, along with hydrogen bond and side chain restraints, as constraints in molecular models of the conformation.

The aggregation state of many peptides has been probed using solution ^{1}H NMR. The method can also be used to determine the degree of ionization of termini or side chains, for example as a function of concentration. Likewise, ^{1}H NMR solubility measurements have been used to obtain solubility phase diagrams and related diffusion measurements have enabled the determination of critical aggregation concentrations (Section 3.12.1). Specific chemical information, for example on metal ion binding to peptides, can also be obtained from solution ^{1}H NMR.

Solid state NMR (ss-NMR) has provided much detail on amyloid fibril structure. A number of high-resolution experiments employing magic angle spinning (MAS) may be performed using isotopically labelled ^{13}C or ^{15}N peptides. MAS is a technique to improve the spectral resolution by averaging orientation-dependent dipolar interactions. Using this method, information on strand registry can be obtained. Homonuclear and heteronuclear 2D and 3D NMR spectroscopy enable information on interatomic distances and torsion angles to be obtained for isotopically labelled peptides in the dried state.

Hydrogen/deuterium exchange techniques have also been employed to probe structural changes during amyloid aggregation. Information on structural rearrangements of subsegments of a protein or peptide during folding, unfolding, or fibrillization can be obtained from exchange rate measurements. This method has also been used, for example, to provide a 3D structure for the amyloid β peptide Aβ42 (Section 3.3).

1.5.5   X-Ray Diffraction

Single-crystal X-ray diffraction (XRD) can be used to obtain the crystal structure of peptides, although there are relatively few reports on this for aggregating peptides due to the difficulty in growing a crystal from a sample in which self-assembly occurs (in preference to crystal formation). In particular, with a very few exceptions, amyloid peptides do not crystallize, prohibiting single-crystal XRD. This is because fibrillar assemblies are by their nature non-crystalline, one- or two-dimensional arrays with a low degree of molecular ordering. Dehydration of the sample to prepare a crystal can disrupt the hydrogen bonding of both α-helical and β-sheet structures. However, with careful preparation, it is possible to obtain single-crystal XRD patterns from small amyloid peptide fragments (Section 3.6).

Fibre diffraction is an alternative to single-crystal XRD that is particularly suited to analyse the structure of peptides which form fibrillar assemblies, including amyloid peptides, coiled coil peptides, lipopeptide fibrils, and peptide nanotubes. Fibre XRD is performed on dried samples, in the form of films or 'stalks', the latter being dried threads of solution. Other methods of alignment include the use of stretch frames or cryo-loops, the latter producing a dried flat film or 'mat'.

1.6   PEPTIDE DATABASES AND WEB SOFTWARE

Many web-based servers are available to calculate simple properties of peptides such as molar mass and pI value, given the input sequence. There are too many of these (and website operability varies) to list. Table 1.6 lists examples of freely downloadable software useful for more detailed peptide property calculation, along with web resources such as databases of different classes of peptides, aggregation propensity calculators, and others.

Table 1.6 Examples of useful peptide software and websites.

Web-server (WS) /stand-alone software (SAS)	Purpose	Website
molinspiration	Useful to calculate molar mass/volume and rule-of-5 parameters, e.g. log$P$ for structures (e.g. peptide conjugates) drawn on web applet	https://www.molinspiration.com/cgi-bin/properties
Pymol (SAS)	Graphics for pdb files and built sequences, H-bond analysis, etc.	https://pymol.org/edu/?q=educational A fuller featured commercial version is also available
Chimera (SAS)	Powerful graphics capability for pdb files and built sequences and various analysis tools including H-bond analysis, solvation, Ramachandran plot generator, and others	https://www.cgl.ucsf.edu/chimera
Uppsala Ramachandran Server	Ramachandran plots (from pdb files)	http://eds.bmc.uu.se/ramachan.html
Peptides package for R (SAS, R is a package for statistical computing)	Various properties including hydropathy analysis, hydrophobic moment calculation	https://cran.r-project.org/web/packages/Peptides/index.html
Transporter classification database	Hydropathy analysis, hydrophobic moment calculation, helical wheel plots, sequence alignment	http://www.tcdb.org/analyze.php
Pepwheel	Helical wheel projection plots	http://www.bioinformatics.nl/cgi-bin/emboss/pepwheel
heliquest	Properties of different helical structures including hydrophobicity, hydrophobic moment, and helical wheel representations	http://heliquest.ipmc.cnrs.fr/cgi-bin/ComputParams.py

Table 1.6 (*continued*)

Web-server (WS) /stand-alone software (SAS)	Purpose	Website
ISAMBARD	Biomolecular design and analysis, particular useful for coiled coils	https://github.com/ isambard-uob/ isambard
CC Builder	Coiled coil peptide modelling and design	http://coiledcoils.chm .bris.ac.uk/ccbuilder2/ builder
LOGICOIL	Predicts oligomerization state of coiled coils	http://coiledcoils.chm .bris.ac.uk/ LOGICOIL
ExPASy	Huge range of tools for coiled coil calculations/M_w and pI calculations, pdb viewing, sequencing, and others (mostly for proteins)	https://www.expasy.org/ structural_ bioinformatics
PepFold	Secondary structure prediction from sequence	http://bioserv.rpbs.univ-paris-diderot.fr/ services/PEP-FOLD3
SOPMA	Secondary structure prediction from sequence	https://npsa-prabi.ibcp .fr/cgi-bin/npsa_ automat.pl? page=npsa_sopma .html
BLAST	Sequence database for alignment checking	https://blast.ncbi.nlm .nih.gov/Blast.cgi? PAGE=Proteins&
Biophython	Python processing for BLAST, ExPASy searches, and analysis, etc.	https://biopython.org
FASTA	Sequence similarity search	www.ebi.ac.uk/Tools/ services/web/ toolresult.ebi? jobId=fasta-I20191004-135659-0538-29792313-p2m
TANGO	Aggregation domain (amyloid) prediction	http://tango.crg.es
Waltz	Aggregation domain (amyloid) prediction	http://waltz.switchlab .org
Camsol and S2D	Aggregation (amyloid) propensity prediction	http://www-mvsoftware .ch.cam.ac.uk

(*continued*)

Table 1.6 (*continued*)

Web-server (WS) /stand-alone software (SAS)	Purpose	Website
Dichroweb	Secondary structure estimation from CD spectra	http://dichroweb.cryst .bbk.ac.uk/html/home .shtml
Bestsel	Secondary structure estimation from CD spectra	http://bestsel.elte.hu/ index.php
δ2D and camcoil	Predict secondary structures from chemical shifts and calculate random coil chemical shifts from sequence	http://www-mvsoftware .ch.cam.ac.uk
Biological Magnetic Resonance Databank	Protein and peptide chemical shift data	http://bmrb.wisc.edu
Bayreuth University TOCSY and COSY amino acid spectra	Amino acid 2D TOCSY and COSY spectra	http://www.bp.uni-bayreuth.de/NMR/ nmr_aminotocsy.html
Protein Data Bank	Peptide structures from XRD, NMR, and electron microscopy	http://rcsb.org
APD2/3	Natural antimicrobial peptide database	http://aps.unmc.edu/AP
CAMP	Antimicrobial peptide database	http://www.camp .bicnirrh.res.in
YADAMP	Antimicrobial peptide database	http://yadamp.unisa.it
DBAASP	Database of antimicrobial peptides and structure prediction (hydrophobicity, tilt angle, pI, etc.)	https://dbaasp.org/home
MilkAMP Database	Milk antimicrobial peptide database	http://milkampdb.org/ home.php
THpdb	Database of FDA-approved therapeutic peptides and proteins	http://crdd.osdd.net/ raghava/thpdb

BIBLIOGRAPHY

Barth, A. (2007). Infrared spectroscopy of proteins. *Biochimica Et Biophysica Acta-Bioenergetics* 1767: 1073–1101.

Branden, C. and Tooze, J. (1999). *Introduction to Protein Structure*. New York: Garland Publishing.

Castelletto, V. and Hamley, I.W. (2018). Methods to characterize the nanostructure and molecular organization of amphiphilic peptide assemblies. In: *Peptide Self-Assembly: Methods and Protocols* (eds. B.L. Nilsson and T.M. Doran), 3–21. Totowa: Humana Press Inc.

Castelletto, V., Hamley, I.W., Seitsonen, J. et al. (2018). Conformation and aggregation of selectively PEGylated and lipidated gastric peptide hormone human PYY_{3-36}. *Biomacromolecules* 19: 4320–4332.

Castelletto, V., Edwards-Gayle, C.J.C., Greco, F. et al. (2019). Self-assembly, tunable hydrogel properties and selective anti-cancer activity of a carnosine-derived lipidated peptide. *ACS Applied Materials & Interfaces* 11: 33573–33580.

Chou, P.Y. and Fasman, G.D. (1974). Prediction of protein conformation. *Biochemistry* 13: 222–245.

Creighton, T.E. (1993). *Proteins. Structures and Molecular Properties*. New York: W.H. Freeman.

Crick, F.H.C. (1953). The packing of α-helices: simple coiled coils. *Acta Crystallographica* 6: 689–697.

Edwards-Gayle, C.J.C., Castelletto, V., Hamley, I.W. et al. (2019). Self-assembly, antimicrobial activity and membrane interactions of arginine-capped peptide bola-amphiphiles. *ACS Applied Bio Materials* 2: 2208–2218.

Eisenberg, D., Weiss, R.M., Terwilliger, T.C., and Wilcox, W. (1982). Hydrophobic moments and protein structure. *Faraday Symposia of the Chemical Society* 17: 109–120.

Hamley, I.W. (2007). Peptide fibrillisation. *Angewandte Chemie, International Edition in English* 46: 8128–8147.

Hutchinson, J.A., Hamley, I.W., Torras, J. et al. (2019). Self-assembly of lipopeptides containing short peptide fragments derived from the gastrointestinal hormone PYY_{3-36}: from micelles to amyloid fibrils. *Journal of Physical Chemistry B* 123: 614–621.

Kelly, S.M., Jess, T.J., and Price, N.C. (2005). How to study proteins by circular dichroism. *Biochimica et Biophysica Acta* 1751: 119–139.

Kyte, J. and Doolittle, R.F. (1982). A simple method for displaying the hydropathic character of a protein. *Journal of Molecular Biology* 157: 105–132.

Langel, U., Cravatt, B.F., Gräslund, A. et al. (2010). *Introduction to Peptides and Proteins*. Boca Raton: CRC Press.

Mielke, S.P. and Krishnan, V.V. (2009). Characterization of protein secondary structure from NMR chemical shifts. *Progress in Nuclear Magnetic Resonance Spectroscopy* 54: 141–165.

Moore, D.S. (1985). Amino acid and peptide net charges: a simple calculational procedure. *Biochemical Education* 13: 10–11.

Nordén, B., Rodger, A., and Dafforn, T.R. (2010). *Linear Dichroism and Circular Dichroism: A Textbook on Polarized-Light Spectroscopy*. Cambridge: RSC.

Phoenix, D.A., Dennison, S.R., and Harris, F. (2013). *Antimicrobial Peptides*. Weinheim, Germany: Wiley-VCH.

Schneider, J.P., Pochan, D.J., Ozbas, B. et al. (2002). Responsive hydrogels from the intramolecular folding and self-assembly of a designed peptide. *Journal of the American Chemical Society* 124: 15030–15037.

Sewald, N. and Jakubke, H.-D. (2002). *Peptides: Chemistry and Biology*. Weinheim: Wiley-VCH.

Stuart, B. (1997). *Biological Applications of Infrared Spectroscopy*. Chichester: Wiley.

Su, J.Y., Hodges, R.S., and Kay, C.M. (1994). Effect of chain length on the formation and stability of synthetic α-helical coiled coils. *Biochemistry* 33: 15501–15510.

Surewicz, W.K. and Mantsch, H.H. (1988). New insight into protein secondary structure from resolution-enhanced infrared spectra. *Biochemica et Biophysica Acta* 952: 115–130.

Van Vranken, D. and Weiss, G. (2013). *Introduction to Bioorganic Chemistry and Chemical Biology*. New York: Garland Science.

Voet, D. and Voet, J.G. (1995). *Biochemistry*. New York: Wiley.

Westermann, J.-C. and Craik, D.J. (2008). NMR in peptide drug development. In: *Peptide-Based Drug Design* (ed. L. Otvos). Totowa, New Jersey: Humana Press.

White, S.H. and Wimley, W.C. (1999). Membrane protein folding and stability: physical principles. *Annual Review of Biophysics and Biomolecular Structure* 28: 319–365.

Wishart, D.S. (2011). Interpreting protein chemical shift data. *Progress in Nuclear Magnetic Resonance Spectroscopy* 58: 62–87.

2

Synthesis

2.1 INTRODUCTION

The main route to synthesize peptides in the laboratory is solid-phase peptide synthesis (SPPS). This was among the first chemistries to be developed for automated synthesis, building on Bruce Merrifield's Nobel Prize winning research in the 1960s. It has been possible to perform these reactions to synthesize shorter peptides using commercially available peptide synthesizers for some time now. This is the subject of Section 2.2 in this chapter, which covers the key reactions, choice of solid-phase resins, protecting-group strategies, and common side reactions. Solution-phase synthesis is briefly mentioned in Section 2.3.

Many strategies are available to prepare longer peptides via convergent synthesis methods or via recombinant expression techniques and these are discussed in Section 2.4.

When screening peptides for different types of activities, it is common to prepare large libraries of peptides and methods to produce peptide libraries are outlined in Section 2.5.

Cyclic peptides have evolved naturally, for example several types of antimicrobial peptides are cyclic (Chapter 4). Inspired by this, synthetic cyclic peptides are of great interest since they can exhibit enhanced stability against degradation *in vivo*, and the steric constraints imposed can be used to enhance the selectivity or specificity of interaction of the peptide with cell receptors. Section 2.6 of this chapter discusses methods to prepare cyclic peptides. Molecules that resemble peptides, termed peptidomimetics, are considered in Section 2.7.

Introduction to Peptide Science, First Edition. Ian W. Hamley.
© 2020 John Wiley & Sons Ltd. Published 2020 by John Wiley & Sons Ltd.

This chapter also covers more recent advances in the synthesis of hybrid peptide biomaterials such as polymer–peptide conjugates. Such methods underpin developments in chemical biology related to peptides and proteins. Section 2.8 introduces post-translational modifications, which include proteolytic processing, modifications of the termini, lipidation, and glycosylation. Conjugation of lipids, glycopolymers (poly- or oligo-saccharides), or polymers can improve the stability and activity of peptides and synthetic methods to attach these moieties are discussed in turn in Sections 2.9–2.11.

Some bioactive natural peptides are synthesized *in vivo* non-ribosomally. Non-ribosomal peptide synthesis (NRPS) is touched on in this chapter in Section 2.12. The last part of this chapter (Section 2.13) discusses the essential methods to purify and characterize a peptide or peptide conjugate after synthesis.

2.2 SOLID-PHASE PEPTIDE SYNTHESIS

2.2.1 General Scheme

The most common method to prepare standard peptides is sequential solid-phase peptide synthesis (SPPS). This method can be automated and is used in commercial peptide synthesizers. These usually operate in batch mode, although continuous flow reactors are also possible. Solid-phase methods offer the advantage that reaction side products can be washed from the resin beads and additionally the method avoids solubility problems encountered in solution-phase synthesis. Disadvantages include the fact that large excesses of amino acids are required, there is a risk of side reactions during the synthesis, monitoring of the reaction process is difficult, and aggregation may hinder the synthesis. SPPS is generally performed from the C-terminus to the N-terminus. Figure 2.1 shows a schematic of the sequential loading, deprotection, activation, and coupling steps, with washing stages in between to remove excess reagents. At the final stage, the peptide is cleaved from the resin.

SPPS methods can be used to prepare peptides typically up to 20–40 residues in length. If the yield for each step is 90% then for a 20-residue peptide the overall yield (with 19 coupling reactions) would be $(0.9)^{19} = 0.14$. The overall yield would be significantly higher say for a 99% yield per step but this is hard to achieve in practice.

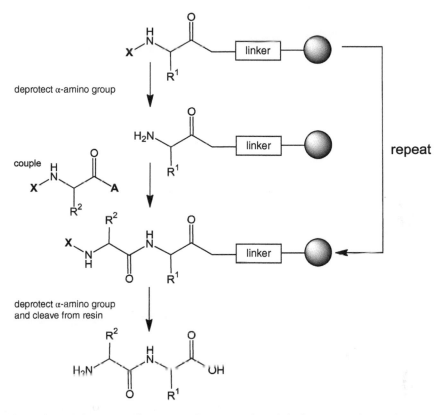

Figure 2.1 Schematic of sequential process in solid-phase peptide synthesis. **X** denotes the amino protecting group and **A** indicates a carboxy activating group (discussed further in Section 2.2.5). The residue groups R^1 and R^2 may also need (orthogonally) protecting if they contain reactive groups.

2.2.2   Resins

SPPS relies on the use of resins as supports for sequential coupling reactions. The resins are 20–80 µm diameter cross-linked polymeric beads (typically polystyrene cross-linked with divinyl benzene) with a surface coating of linker groups for the coupling of the first amino acid. The structures of commonly used SPPS resins are shown in Table 2.1. The linkers are designed to be stable during the peptide synthesis reactions, but to be cleavable at the final stage of removal of the peptide from the support. The chloromethyl resin originally developed by Merrifield is not widely used nowadays.

Table 2.1 Common peptide synthesis resins (resin bead shown as filled circle).

Resin name	Resin structure	Cleavage conditions	Peptide product
Wang resin	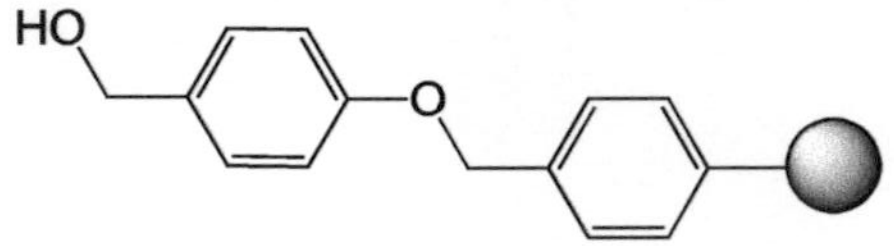	90–90% TFA in CH_2Cl_2, 1–2 h	Acid
Rink amide resin	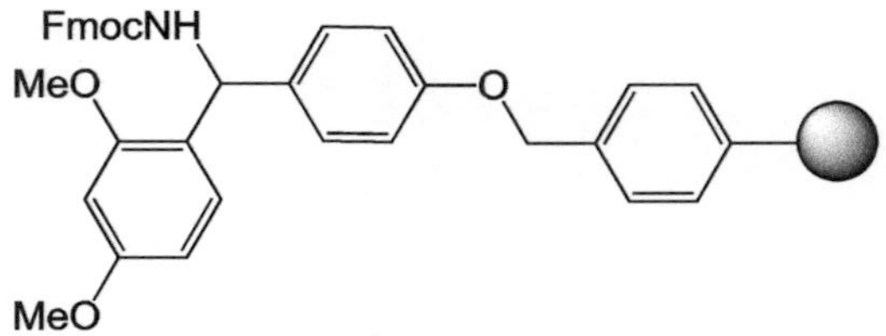	50% TFA in CH_2Cl_2, 1 h	Amide
Rink acid resin	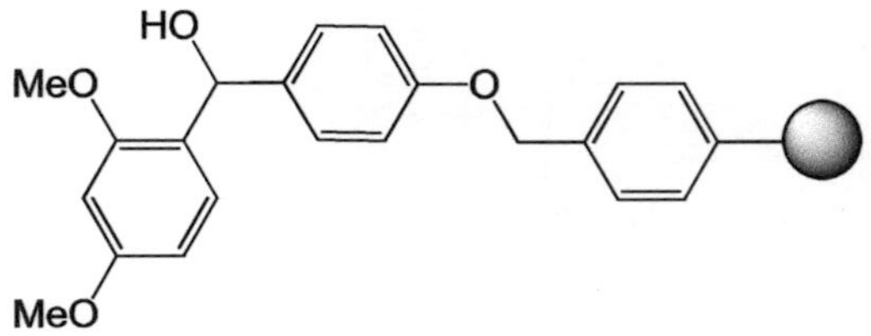	1–5% TFA in CH_2Cl_2, 5–15 min or 10% acetic acid in CH_2Cl_2	Acid
MBHA (4-methyl benzhydrylamine) resin	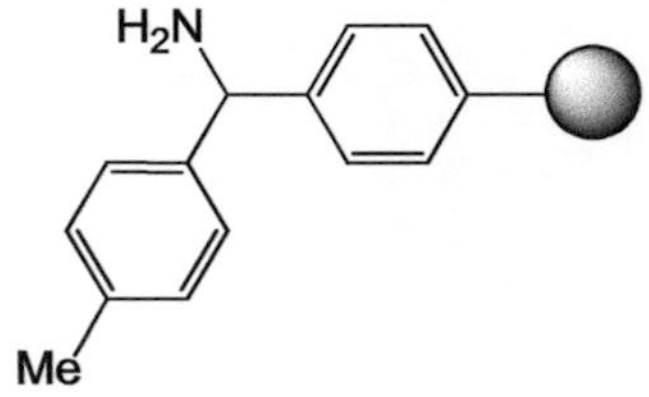	HF 0 °C, 1 h	Amide

HMPB [(4-hydroxymethyl-
3-methoxyphenoxy) butanoic
acid] resin

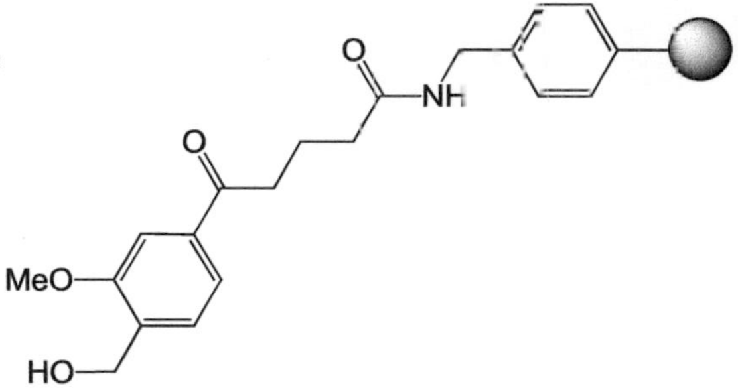

1% TFA in CH$_2$Cl$_2$, 2–5 min

Acid

2-chlorotrityl chloride resin

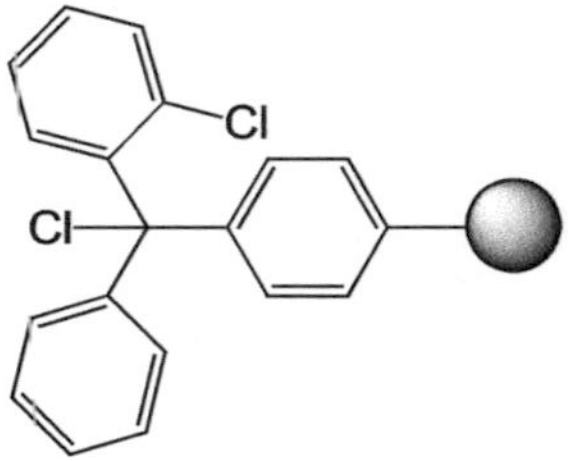

1–5% TFA in CH$_2$Cl$_2$, 1 min

Acid

SASrin resin

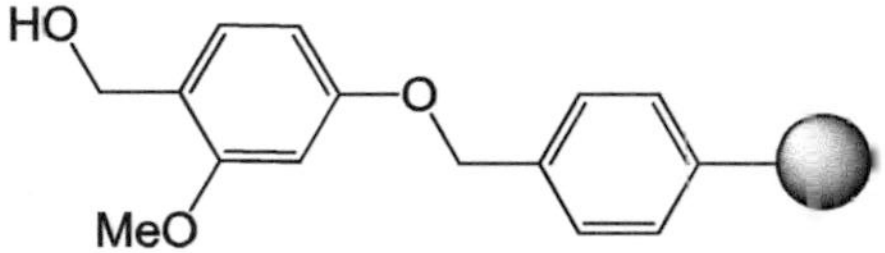

1% TFA in CH$_2$Cl$_2$, 5–10 min

Acid

HMBA (4-hydroxymethyl
benzoic acid) resin

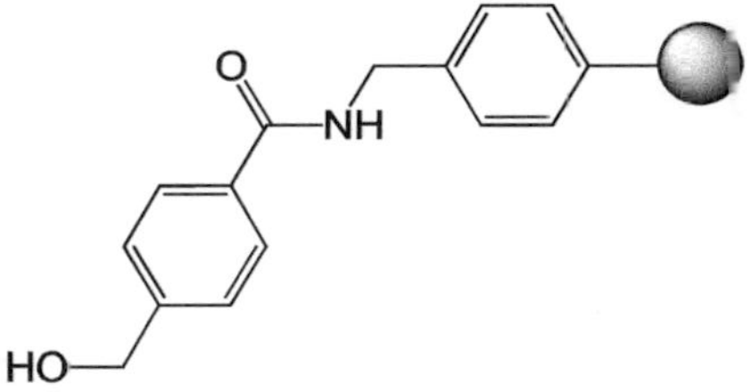

NaOH, N$_2$H$_4$, N$_2$ in MeOH 24 h

Acid,
hydrazide,
amide

ROH
LiB$_4$

Ester
Alcohol

2.2.3 Protecting Groups

A number of protecting group chemistries are available for the N-terminus, C-terminus, and side chains. So-called orthogonal protecting groups are usually employed so that termini and side groups can be deprotected separately. The most widely used protecting group chemistry is Fmoc/tBu, although the earlier Merrifield Boc/Bzl strategy may also be used. Here, Fmoc denotes fluorenylmethyloxycarbonyl, tBu denotes *tert*-butyl, Boc indicates butyloxycarbonyl, and Bzl indicates benzyl. The structures of these protecting groups are shown in Figure 2.2. The C-terminus is generally protected on the resin and the Fmoc or Boc units protect the N-terminus. When protection of the C-terminal carboxy group is required, esterification may be performed using methyl, ether, *tert*-butyl, benzyl esters, etc. Fmoc is stable in acidic conditions, and is removed in piperidine solution. Boc is removed in TFA (trifluoroacetic acid) solution and is stable in basic conditions.

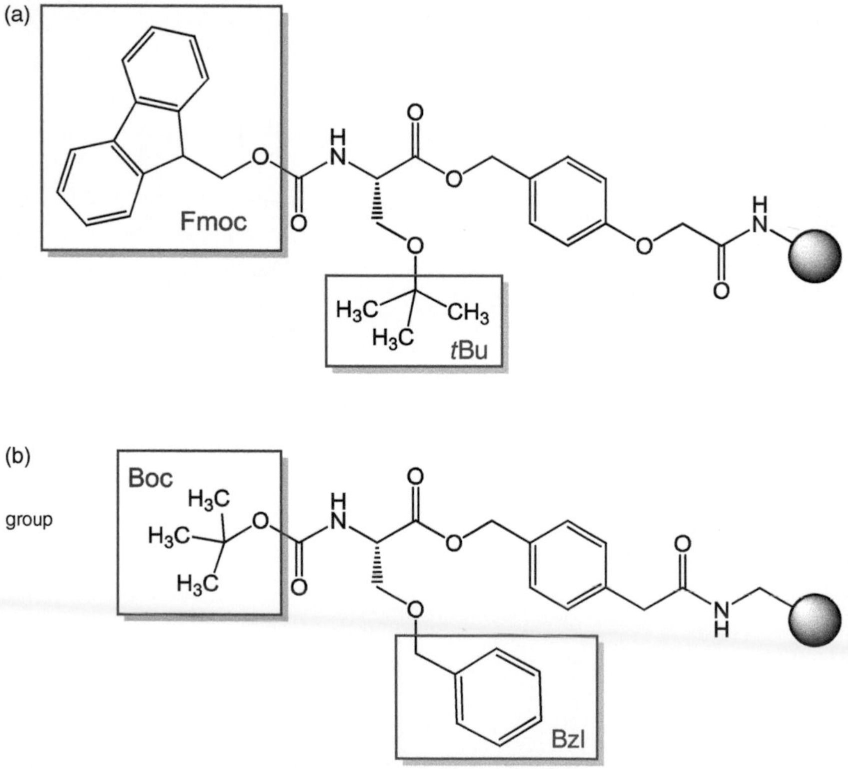

Figure 2.2 Orthogonal protecting groups illustrated for an aspartic acid residue on resin. (a) Fmoc/tBu protection, (b) Boc/Bzl protection.

There are a considerable range of known side chain protecting groups, examples of which are listed in Table 2.2. The optimum cleavage conditions are also listed; often this is done in a TFA solution (90–95%) in dichloromethane (DCM), although some side groups require other deprotection chemistries.

2.2.4 Deprotection

In the case of the standard Fmoc protocol, the Fmoc group on each amino acid is removed using 20–50% piperidine in DMF (dimethyl formamide) or NMP (N-methyl pyrollidone).

2.2.5 Activation/Coupling

The most common chemistry for SPPS involves the use of Fmoc chemistry with side chain protected amino acids. Common reagents for activation of the carboxyl group during the coupling stage are shown in Figure 2.3. DCC (N,N′-dicyclohexylcarbodiimide) can be used in solution-phase synthesis but leads to the insoluble by-product dicyclohexylurea, therefore DIC (N,N′-diisopropylcarbodiimide) was introduced for solid-phase synthesis since the urea by-product remains soluble. Later, additional reagents, especially HOBt (hydroxybenzotriazole), were developed (usually used along with DIC) to produce active esters, suppressing racemization. Derivatives of HOBt were subsequently developed to generate active esters without the need for carbodiimides at all.

The phosphonium reagent PyBOP is widely used. Uronium reagents such as HATU (hexafluorophosphate azabenzotriazole tetramethyl uronium) and HBTU (hexafluorophosphate benzotriazole tetramethyl uronium) are also extensively employed (HATU reacts more quickly than HBTU with a lower degree of epimerization). The mechanism of the reaction which produces OBt (benzotriazole) esters is illustrated for HBTU in Figure 2.4. The carboxyl group of the amino acid attacks the imide carbonyl carbon of HBTU. Then the displaced OBt⁻ attacks the acid carbonyl, giving a tetramethyl urea by-product and the activated amino acid ester. Subsequent aminolysis displaces the OBt to form the desired amide. These reagents are used along with the base DIPEA (N,N-diisopropylethylamine) in DMF. HATU or HBTU are often used along with HOBt, since the combination reduces racemization.

Table 2.2 Typical side chain protecting groups and deprotection conditions.

Amino acid	Protecting group	Structure	Deprotection conditions
R	Pmc (2,2,5,7,8-pentamethyl-chroman-6-sulfonyl)	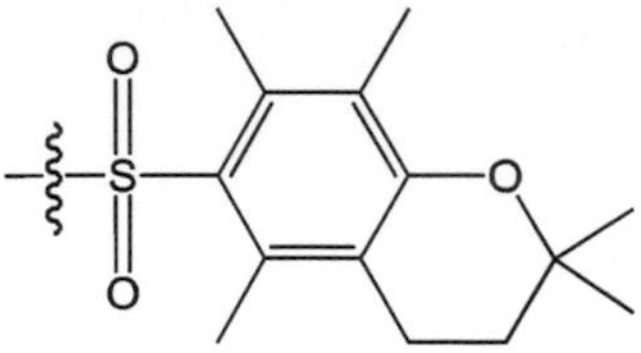	TFA in DCM or in anisole
	Pbf (2,2,4,6,7-pentamethyl-dihydrobenzo-furan-5-sulfonyl)	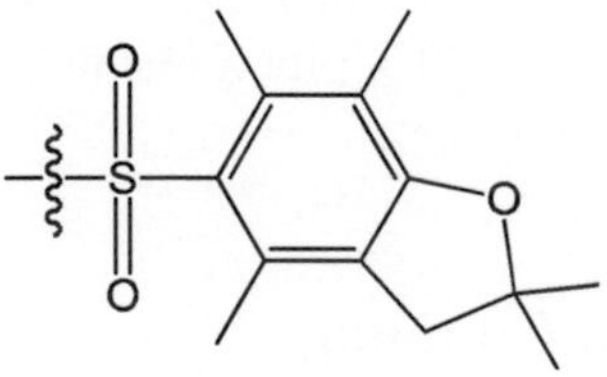	TFA in DCM
D/E	OtBu (tert-butyloxy)	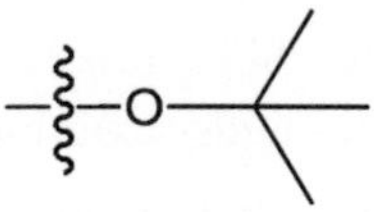	TFA in DCM
	Alloc (allyloxycar-bonyl)	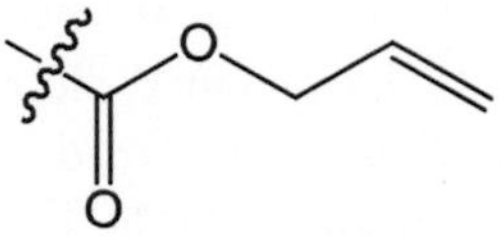	Pd(Ph$_3$P$_4$), PhSiH$_3$ in DCM or THF/CH$_3$OH (THF is tetrahydrofuran)
N/Q	Trt (trityl)	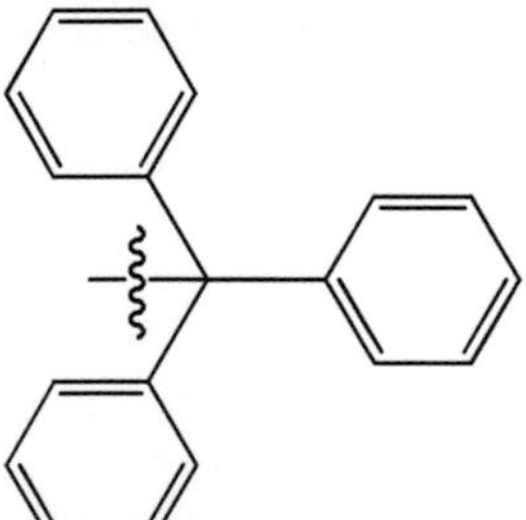	TFA in DCM
	Mtt (4-methyltrityl)	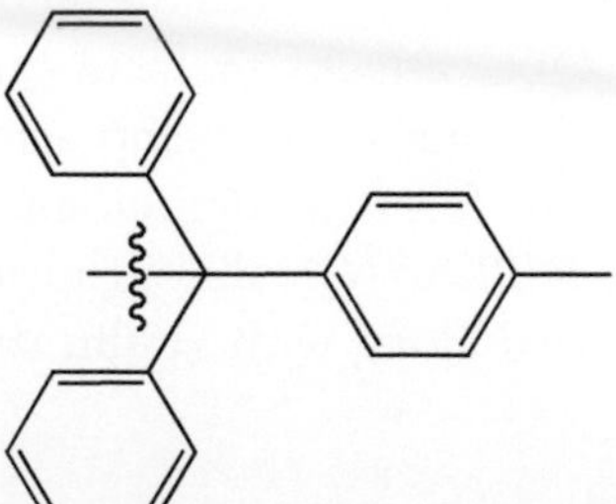	TFA in DCM

Table 2.2 (Continued)

Amino acid	Protecting group	Structure	Deprotection conditions
H	Trt	As above	As above
	Mtt	As above	As above
C	Trt	As above	As above
	Acm (acetamidomethyl)	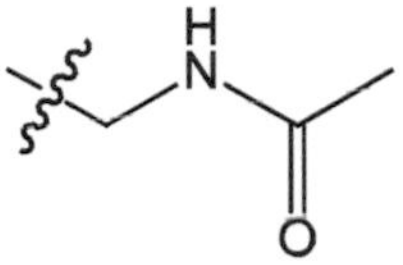	I_2, DTNP [2,2′-dithiobis(5-nitropyridine)], Tl(III), Hg(II)
K	Boc (*tert*-butyloxycarbonyl)	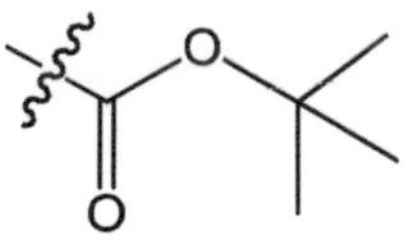	TFA in DCM
	Alloc	As above	As above
S/T/Y	*t*Bu	As above	As above
	Trt	As above	As above
W	Boc	As above	As above
	For (formyl)	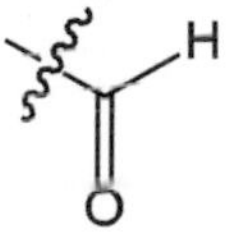	HF in piperidine

Racemization can occur during the coupling stage. The term 'racemization' refers to the loss of enantiomeric purity, i.e. to the formation of a mixture of peptides with L- and D-amino acids. This can result from several mechanisms, including acid- and base-catalysed direct enolization or oxazolone formation, shown schematically in Figure 2.5. Direct enolization follows on from deprotonation of the C_α hydrogen atom. Racemization can be monitored through analysis of the diastereometric purity of the peptide products, which is performed by hydrolysis of the peptide and analysis of the amino acids via chiral chromatography.

2.2.6 Side Reactions and Difficult Sequences

Many possible side reactions can be avoided by appropriate choice of protecting groups. Two commonly observed side reactions are aspartimide formation and diketopiperazine formation. Aspartimide formation may occur for aspartic acid or asparagine residues via the mechanism shown in Figure 2.6 in which the nitrogen on the α-carboxy group attacks an

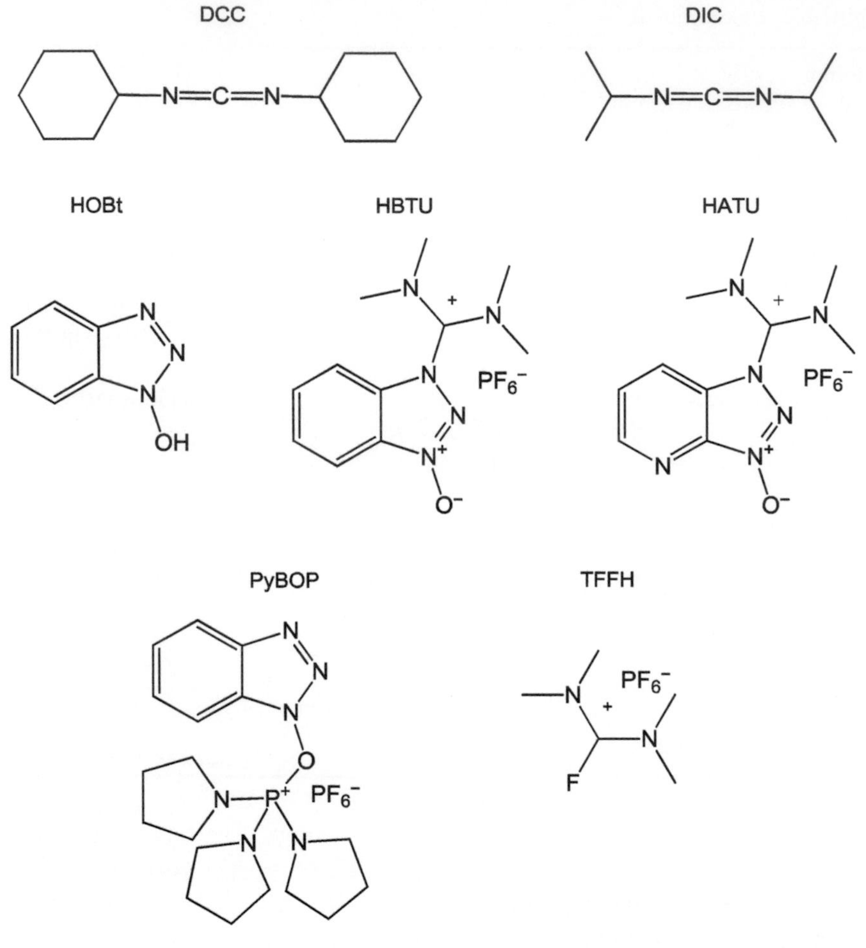

Figure 2.3 Common coupling reagents in peptide synthesis that generate active esters.

ester or amide respectively, leading to the formation of a five-membered imide ring which can undergo further reactions such as hydrolysis to form α- and β-aspartyl peptides or reactions with piperidine. This can even occur with O*t*Bu-protected D residues. The problem can be avoided by protection of the amide nitrogen or in the case of asparagine by protection of the carboxamide. N-terminal glutamine residues can form pyroglutamate (Table 1.3) and since this reaction is catalysed by acids and is problematic using Boc chemistry, it can be avoided using high TFA concentrations for the Boc cleavage.

   Diketopiperazine formation can be problematic for peptides containing proline or *N*-alkylated residues (for example in peptoids) in the two residues at the C-terminus via the mechanism shown in Figure 2.7.

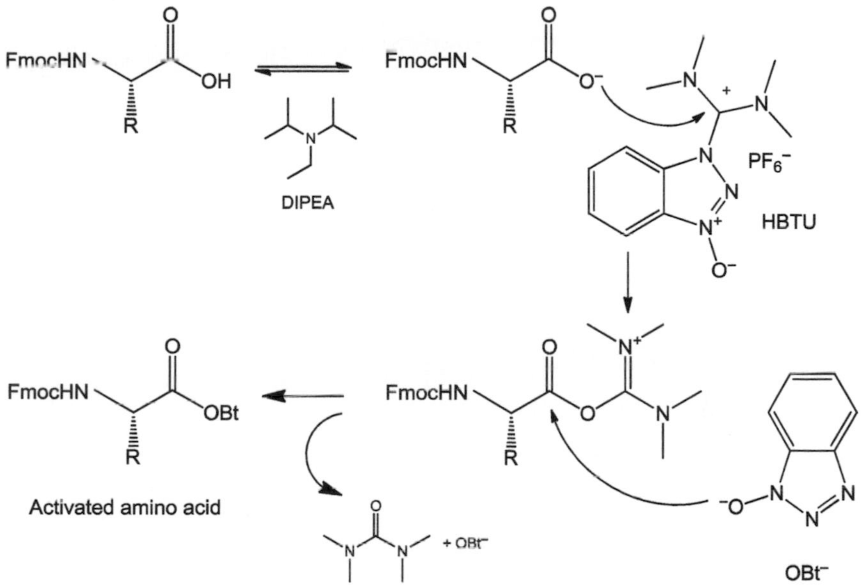

Figure 2.4 HBTU activation mechanism.

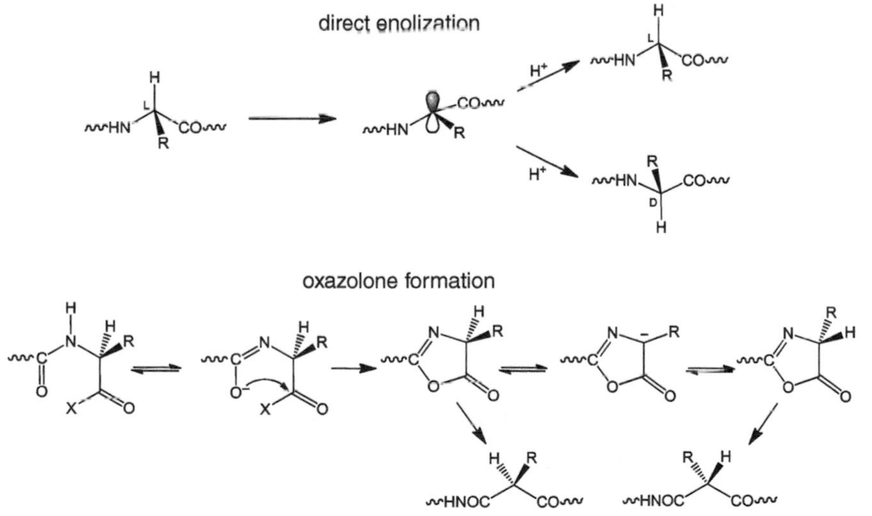

Figure 2.5 Enolization and oxazolone formation cause peptide racemization. X denotes a leaving group.

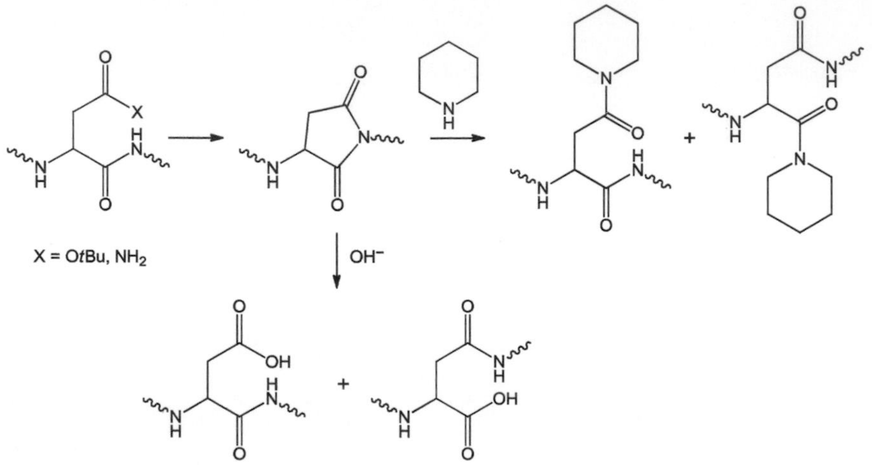

Figure 2.6 Aspartimide formation reaction and subsequent reactions including hydrolysis and reaction with piperidine.

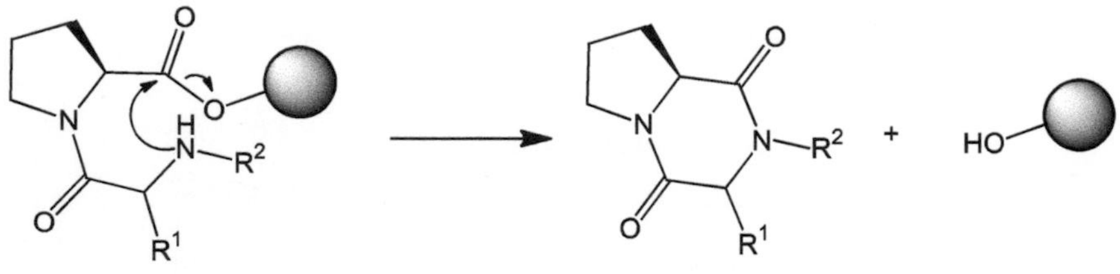

Figure 2.7 Diketopiperazine formation for an on-resin sequence with C-terminal proline, via the indicated cyclization mechanism.

'Difficult sequences' is a term used in the peptide synthesis field that refers to peptide sequences prone to undergo aggregation (β-sheet formation) which hinders the acylation reactions. Random amino acylation can occur with sterically hindered amino acids such those with bulky protecting groups. The synthesis of many long peptides can also be considered 'difficult'. Incorporation of proline residues is a strategy to avoid β-sheet formation but obviously may not be a desirable alteration of the sequence. Pseudoproline formation is another strategy similar to use of prolines. Incorporation of a pseudoproline involves formation of an oxazolidine dipeptide. The dipeptide containing a native serine or threonine (or cysteine) residue is regenerated via treatment of the oxazolidine with TFA (Figure 2.8).

Another strategy to prevent aggregation is to introduce the N-(2-hydroxy-4-methoxybenzyl) (Hmb) group as a temporary backbone protecting group

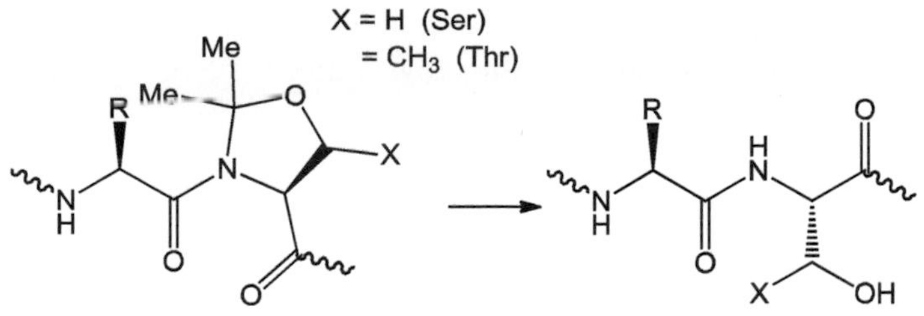

Figure 2.8 Pseudoproline formation strategy: ring-opening of an oxazolidine to produce a serine or threonine residue.

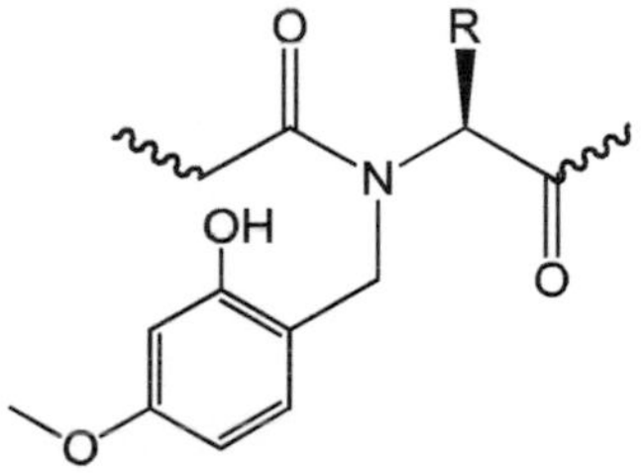

Figure 2.9 The Hmb reversible backbone protecting group.

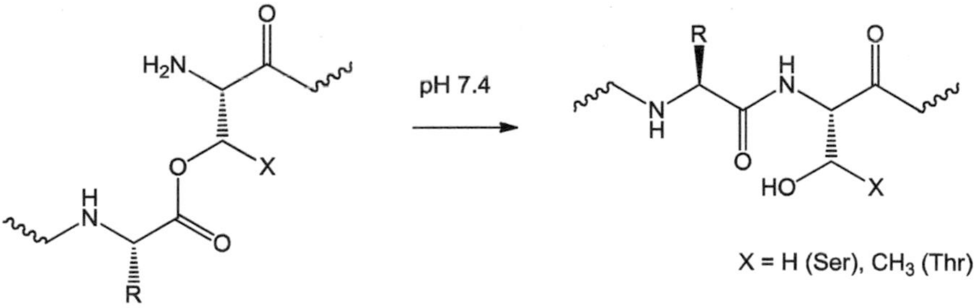

Figure 2.10 O-acyl isopeptide conversion into a native peptide.

(Figure 2.9). This can disrupt the intermolecular hydrogen bonding that characterizes β-sheet aggregates. The Hmb group is acid labile and is removed in the final cleavage step.

A further approach involves the incorporation of an O-acyl isopeptide or depsipeptide group that is more soluble. This is converted (at pH 7.4) via O-N acyl transfer into the native peptide with a Ser or Thr residue (Figure 2.10).

Use of PEGylated resins improves the solubility of peptides during synthesis, which can reduce aggregation. Choice of appropriate solvents that break up aggregates (of stable secondary structures), such as HFIP (hexafluroisopropanol), DMSO (dimethyl sulfoxide), or aprotic solvents are other possible approaches.

2.2.7   Cleavage

Cleavage is usually performed by treating with a 95% solution of TFA for 1–3 h. This removes N-terminal groups such as Fmoc, and also *tert*-butyl protecting groups and certain side chain protecting groups such as trityl (Trt). Due to the use of TFA in the cleavage step, the peptide will be produced as a TFA salt. Ion-exchange methods can be used to remove or change the salt, using preparative HPLC for example.

2.2.8   N-Terminal and C-Terminal Protection

Terminal modifications reduce the charge on peptides, and can enhance the circulation properties of peptides due to protection from exopeptidases (aminopeptidases and carboxypeptidases for the N- and C-termini). N-terminal acetylation can be performed using SPPS methods with acetic acid coupling. C-terminal amidation is achieved via aminolysis of resin-bound esters. Peptide amides can also be prepared using amine-based linkers, as shown in Figure 2.24.

2.2.9   Microwave Synthesis

Many commercial peptide synthesizers now use microwave heating methods to accelerate reactions, such that coupling and deprotection steps only take a few minutes. The common solvents NMP and DMF used in peptide synthesis have a good capacity to absorb microwave radiation.

2.3   SOLUTION-PHASE PEPTIDE SYNTHESIS

Short peptides may be produced manually using sequential solution-phase methods, although this requires considerable handling and is usually only performed for very short peptides. Many of the chemistries employed are similar to those discussed above, although as mentioned the choice of coupling reagents may differ.

2.4 METHODS TO PREPARE LONGER PEPTIDES

2.4.1 Prevention of Aggregation

Methods discussed in Section 2.2.6 for 'difficult sequences' may be used to prevent aggregation during the synthesis of long peptides.

2.4.2 Convergent Synthesis

Fragment condensation methods also known as convergent synthesis methods may be used to link short peptide fragments (typically up to 15 residues) which are easier to produce. Coupling can also produce larger yields, since the yield of a short peptide obtained by SPPS, for example, is (by a power law factor) higher than that of a long peptide. Fragment condensation is beneficial in the synthesis of peptides where solubility becomes problematic, or those containing β-turns which are also considered 'difficult'. Also, since the product is much larger than the fragments, isolation is easier than in the case of extended step-wise synthesis, where after each step the new product is only fractionally different to the previous one. To minimize racemization during the coupling reaction, C-terminal G or P residues are preferred. Since peptides can be synthesized and coupled in a multitude of ways, careful consideration is required for the synthesis strategy for longer peptides.

A number of condensation reactions available are for solution synthesis, including the Sakakibara approach using protected fragments, or for minimally protected peptides, chemical coupling reactions such as the thiocarboxy coupling reaction are available. Chemoenzymatic approaches may also be employed to couple fragments, by generating a C-terminal peptide ester from one fragment that can be coupled to another fragment (with protected N-terminus but uncapped C-terminus) using proteases such as subtilisin, thermolysin, or trypsin via reverse hydrolysis (cf. Figure 3.15).

Convergent SPPS is a convenient method to couple fragments on resin. Figure 2.11 shows the sequential reactions schematically. The assembly can be performed from C-terminus to N-terminus or the reverse, although the former is preferred.

Native chemical ligation (NCL) is a convenient method to link peptide fragments bearing C-terminal thioesters and N-terminal cysteine residues. This reaction is performed under mild aqueous conditions. Figure 2.12 shows the mechanism. The Staudinger ligation reaction, discussed further in Section 2.11.2, has also been used to couple peptide fragments, avoiding the need for a terminal cysteine residue on one fragment. Other methods have

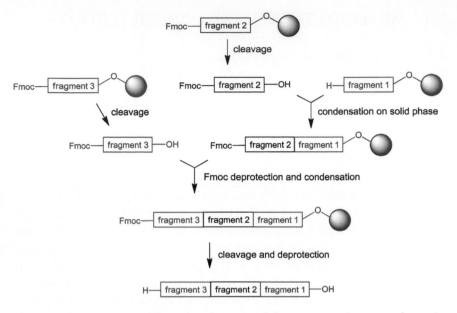

Figure 2.11 Convergent SPPS, for the case of fragment condensation from the C-terminus to the N-terminus.

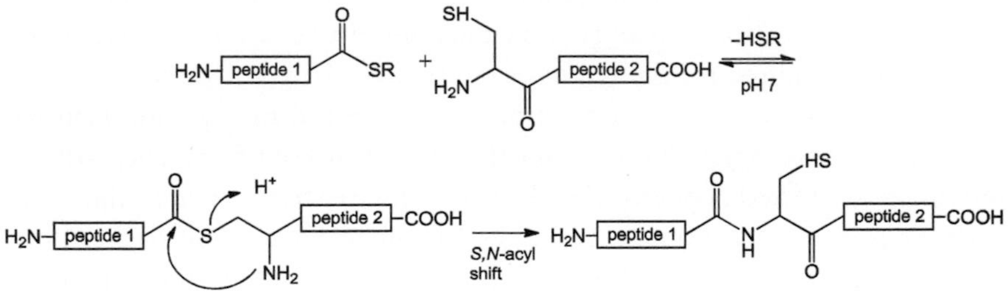

Figure 2.12 Mechanism of native chemical ligation, here the SR group is present in a thioester.

been introduced recently, including ketoacid-hydroxylamine and thioester ligation reactions.

2.4.3 Recombinant Methods

Many non-aggregating peptides can be expressed by genetic engineering using recombinant DNA methods (molecular cloning) with natural DNA sequences, or using synthetic DNA. This method is well suited

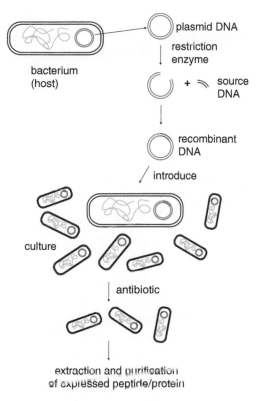

Figure 2.13 Recombinant production of peptides/proteins via molecular cloning using plasmid DNA vectors.

to expression of longer peptides and proteins, and can be scaled up. Figure 2.13 summarizes the process. First, synthetic DNA corresponding to the peptide of interest or complementary DNA (cDNA) obtained from mRNA of a clone of interest is obtained. A host is selected, which is typically a bacterium since bacteria can be easily cultivated. The host vector DNA is prepared using a restriction enzyme (restriction endonuclease) to cleave the DNA at the site where the DNA fragment of interest is to be inserted (this is also treated with restriction endonuclease in preparation for insertion). This fragment is then spliced (ligated) into the DNA of the host organism using the enzyme DNA ligase. This is usually done using plasmid DNA (which forms a loop) from a bacterium, especially *Escherichia coli*. Plasmids containing antibiotic resistance genes are often used as selectable markers, since during the culturing process those that do not contain this gene can be killed by the antibiotic (e.g. ampicillin). The recombinant DNA is then introduced to the host organism via transformation, transfection, or transduction. Transformation refers to the process in which competent microorganisms take up the exogenous DNA from their surroundings through their cell membranes. Transfection is the term used for this process during the culturing of mammalian cells. Transduction involves the use of viruses to transfer the DNA into the host cells. The host cells (typically bacteria) are then cultivated and the expressed recombinant peptide or protein is then extracted and purified.

Biosynthetic insulin (Section 5.3.3) is an example of an important therapeutic peptide produced by recombinant DNA genetic engineering methods. Human growth hormone is another medically important hormone (discussed further in Section 5.3.1) produced in this way.

2.5 PEPTIDE LIBRARY SYNTHESIS

Parallel methods enable multiple peptide synthesis through combinatorial techniques. A number of techniques have been developed including teabag or T-bag manual synthesis and automated (or automatable) spot synthesis, multipin synthesis, and other methods. The teabag method can be used to produce typical 10–50 mg quantities of each peptide within a library. It uses labelled polypropylene bags containing the resin beads. The bag has a pore size that enables diffusion of reagents in, but the beads are retained. The number of teabags corresponds to the number of peptides to be synthesized. Coupling reactions are performed using standard SPPS methods. After each step, the teabags are pooled and protecting groups removed before sorting and appropriate recombination. The method has the advantage that washing and coupling reactions (and final cleavage) are performed on all the bags together.

Spot and multipin synthesis are suitable for preparing small amounts of peptides and can be automated. The spot method creates arrays on cellulose strips (filter paper). The surface is first functionalized with amine groups, typically via reaction of Fmoc-β-Ala-OH with the cellulose hydroxyl groups, followed by deprotection. The next Fmoc amino acid (or linker) is then coupled using a pipetting robot (or by hand) to distribute small aliquots (typically 1 µl) in a square array (Figure 2.14). Incorporation of linkers may be beneficial when wishing to perform biological assays directly on the peptide array on the strip, since they facilitate contact with (large) biomolecules. Washing and cleavage steps may then be performed on the whole array by dipping the cellulose sheet into appropriate reagent solutions. This method is used to produce arrays containing from 96 up to thousands of peptides. Spot synthesis can be performed using commercial instruments.

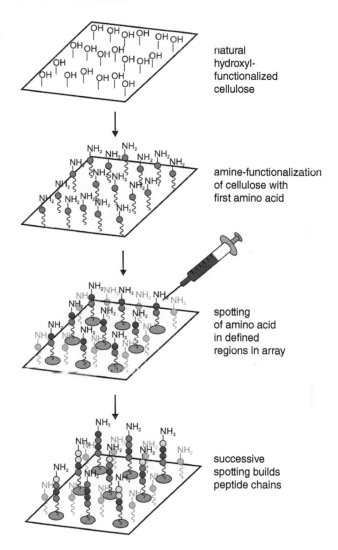

natural
hydroxyl-
functionalized
cellulose

amine-functionalization
of cellulose with
first amino acid

spotting
of amino acid
in defined
regions in array

successive
spotting builds
peptide chains

Figure 2.14 Schematic showing the steps of spot peptide synthesis.

The multipin method is similar, but uses an array of polyethylene pins for the synthesis. The pins are coated with Fmoc-protected linker moieties for the synthesis, and the coupling reactions are performed in parallel in the wells of a microtiter plate. Cleavage and washing steps can be performed for the whole array at the same time.

Mixed peptide libraries can be prepared by combinatorial methods. The split-mix method involves splitting a pool of monomers, coupling, and mixing and then repeating this process sequentially to the desired peptide length.

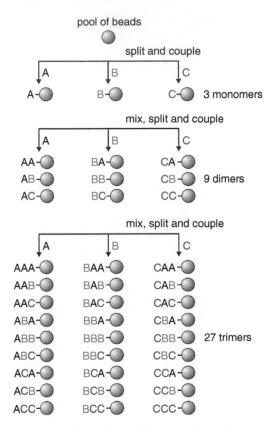

Figure 2.15 Split-mix peptide synthesis, illustrated for a three monomer synthesis.

It is a conceptually simple method, as illustrated in Figure 2.15 for the case of a synthesis using three amino acids.

The simpler reagent mixture method, as the name suggests, just uses a mixture of reagents in large excess to incorporate different residues at a given substitution position. This method is complicated since equimolar addition of the different amino acids can only be achieved with detailed knowledge of the reactivity of each one.

However they have been prepared, the library of synthesized peptides may be screened to identify the active peptides (for example by analysis of binding or biological activity). The split-mix method and other peptide library synthesis techniques require subsequent analysis of the active peptides from the library, which can be a laborious process (using methods described in Section 2.13). This can be overcome by encoding the amino acids using attached tags, which are detachable groups defined to be readily screened by electrophoretic methods. Alternatively, the library may be deconvoluted into a set of smaller sub-libraries.

Phage display is becoming increasingly used as a method to prepare arrays of peptides (or proteins) for screening studies (so-called 'biopanning', i.e. analysis of binding to targets) by genetic engineering of the coat proteins of bacteriophages to produce large random peptide libraries. A phage or bacteriophage is a virus that infects bacteria. The peptides are then displayed on the surface of the phages. Phage display enables peptides to be screened in a process analogous to natural selection, by isolating the first successful displayed peptides, then creating mutants of these, and performing a further round of phase display 'evolution' and so on.

2.6 SYNTHESIS OF CYCLIC PEPTIDES

Reactions to form cyclic peptides can be classified into those involving back bone cyclization (head-to tail cyclization), those involving side chain closure with the peptide head or tail, or side chain to side chain linking reactions (Figure 2.16).

The simplest head-to-tail cyclization is that of diketopiperazines. Although mentioned in Section 2.2.6 as unwanted side products, these species may be of interest for drug discovery. They are easily formed via aminolysis of dipeptide esters. Longer peptides can also be cyclized, either in solution or on resin. It should be considered, however, that there are conformational constraints on ring formation in large organic molecules, i.e. not all cyclizations are feasible sterically/stereochemically. Head-to-tail

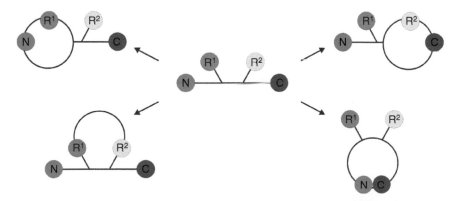

Figure 2.16 Schematic of cyclization reactions involving the peptide head (C-terminus, red) or tail (N-terminus, blue) or side chains R^1, R^2 (green, yellow).

cyclization reactions in solution should be performed under conditions of high dilution to avoid intermolecular dimerization. Standard coupling reactions of activated precursors (for example, activated esters) may be employed. Cyclo-oligomerisation reactions may also be performed to couple dipeptide methyl esters using metal ions or enzymes. Enzymes may also be used to directly couple N- and C-termini to form closed rings. In the solid-phase method, the cyclization may be performed in solution after cleavage. On-resin cyclization is achieved using a resin-bound side chain (Figure 2.17) with orthogonal protection of the C-terminal carboxy group. Thiol chemistry may also be used to achieve backbone cyclization via disulfide bridge formation or using the NCL approach for a peptide bearing an N-terminal cysteine and a C-terminal thioester. NCL is discussed further in Section 2.4.2. The Staudinger ligation discussed in Section 2.11.2 may also be employed to make cyclic head-to-tail peptides.

Side chain-to-head and side chain-to-tail cyclizations have been achieved using a range of chemistries, including lactam formation between a side chain amino group and the C-terminus or between a side chain carboxy group and the N-terminus (a lactam is a cyclic amide). Thioalkylation can also be performed and on-resin cyclizations are possible.

Side chain-to-side chain cyclization can be achieved via many reactions (Figure 2.18), including the formation of disulfide bridges, thioethers, or lactams. Ring-closing metathesis reactions are also possible using modified amino acids bearing alkene (olefin) or alkyne groups, as are click reactions such as the alkyne-azide or thiol-ene reactions. These chemistries are also used for so-called stapling reactions, which are used to stabilize helical

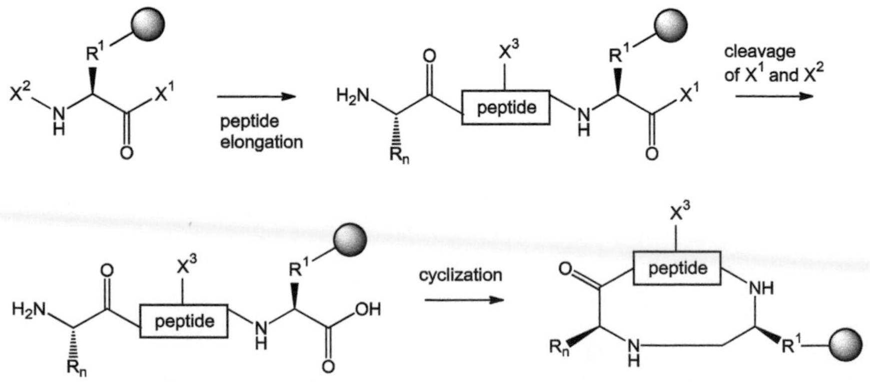

Figure 2.17 Strategy for on-resin head-to-tail peptide cyclization using a resin-bound side chain R^1. Here, X^1 = carboxy protecting group, X^2 = temporary amino protecting group, X^3 = semipermanent side chain protecting groups.

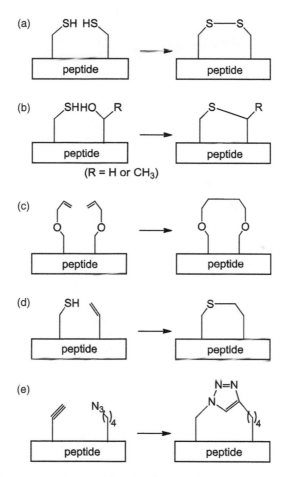

Figure 2.18 Representative peptide side chain cyclization reactions. (a) Disulfide bridge formation, (b) thioether formation via Michael addition, (c) olefin metathesis, (d) thiol-ene click reaction, (e) alkyl-azide click reaction (illustrated for a lysine-derived azide).

peptide conformations. This technique is of great interest in the development of therapeutic peptides able to modulate protein–protein interactions (PPIs, Section 5.7).

Recent interest in cyclic peptides with enhanced biostability has led to the development of bicyclic peptides, with potential interest as therapeutics. The conformational constraints imposed by the two interlocking rings lead to molecules with the potential to exhibit high target specificity and selectivity. Bicyclic peptides occur in nature, for example they are produced by fungi as toxins (Phalloidin and α-amanitin are examples – see Figure 2.19). Bicyclic

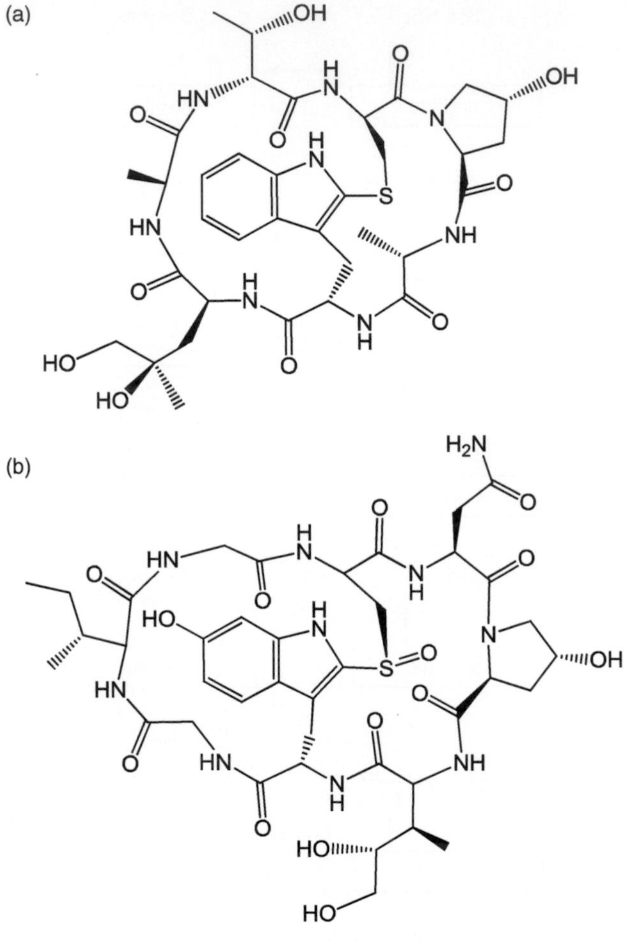

Figure 2.19 Molecular structures of two natural bicyclic peptides: (a) phalloidin, (b) α-amanitin.

peptides are also on the market as peptide therapeutics; some examples are discussed in Section 5.6.

Many approaches detailed in the preceding paragraphs can be extended to synthesize bicyclic peptides. The simplest motif is a head-to-tail cyclized peptide with an additional side chain-to-side chain linkage such as a disulfide bridge, although other architectures including that of a linear peptide cyclized through two side chain to side chain linkers are possible. As well as disulfide links, olefin or alkyne ring closing (cross-) metathesis, thioether or ether cross-linking reactions may be performed (Figure 2.18). For cross-linking of linear peptides, two cross-linking reactions may be of the same type or they may be different. Orthogonal reactions may also be employed.

Multiply cyclized peptides are also found in nature; examples are the conotoxins discussed further in Section 5.5. Many of these contain multiple disulfide linkages.

2.7 PEPTIDOMIMETICS

Peptidomimetics (also known as pseudopeptides) are of interest in the development of analogues of peptides that are more stable *in vivo* due to reduced (or eliminated) proteolysis. Figure 2.20 shows examples of peptidomimetic structures. Depsipeptides are peptide analogues with an ester backbone link rather than amide as in a peptide, which provides resistance against endopeptidases. This resistance is also a feature of β- and γ-peptides. Several classes of natural antimicrobial peptides are depsipeptides. Peptoids are also natural products; they are analogues of peptides with substituents located on the backbone nitrogen instead of the C_α atom. These groups include many analogues of those found in amino acids. Peptoids are N-substituted

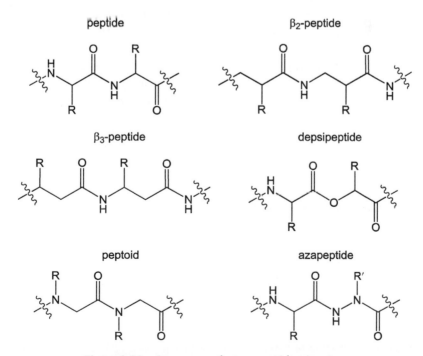

Figure 2.20 Structures of some peptidomimetics.

glycines. Peptoids have non-chiral backbones. Azapeptides have conformational restriction due to the aza group, which favours turn conformations. In addition to the other peptidomimetics in Figure 2.20, a variety of 'foldamer' structures with different backbones have been synthesized and have been shown to fold into structures reminiscent of peptides/proteins, for example adopting a range of helical conformations. These structures are stabilized by non-covalent interactions such as π-stacking interactions.

Another class of peptidomimetics are peptide nucleic acids (PNAs), which comprise a pseudopeptide backbone with pendant nucleobase units. These molecules are of potential interest in the development of gene therapies since PNAs can interact with DNA and RNA.

2.8 POST-TRANSLATIONAL MODIFICATIONS

In nature, various modifications of peptides (and proteins) have evolved that improve their activity or stability. These modifications are made post-translationally, i.e. after the ribosomal synthesis of the peptide (protein). Proteolytic cleavage occurs for many families of peptide hormones, as discussed in Section 5.3. Examples are insulin (from preproinsulin) and the vasopressins (Section 5.3.4). Modifications at the peptide terminus are also carried out post-translationally. For example, N-terminal acetylation is common and is performed by N^α-acetyltransferases. Attachment of N-terminal fatty acids (lipidation) via laboratory synthesis is discussed in Section 2.9. At the C-terminus, the main modification is attachment of a glycosylphosphatidylinositol (GPI) anchor, which is a membrane-anchoring group that is a complex glycolipid. C-terminal amidation is another common modification, especially for peptide hormones. The amide terminus is usually derived from a C-terminal glycine residue via oxygenation in the presence of the enzyme peptidyl-glycine-α-amidating monooxygenase (Figure 2.21). Prenylation is a C-terminal modification that can occur at a C-terminal cysteine residue, as discussed in Section 2.9.

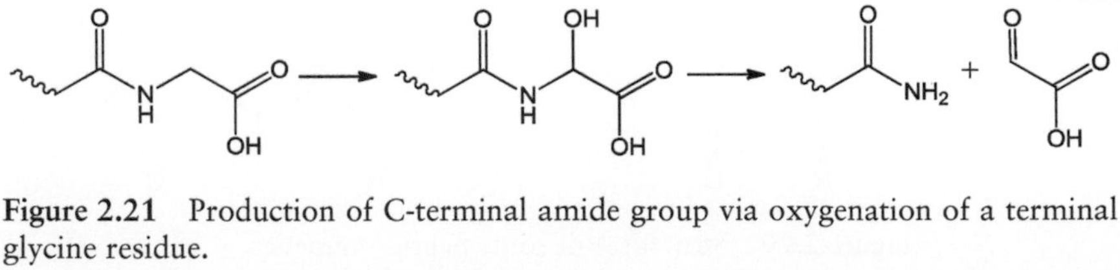

Figure 2.21 Production of C-terminal amide group via oxygenation of a terminal glycine residue.

Phosphorylation is a common post-translational modification which is carried out by kinases using ATP (adenosine triphosphate) as a phosphate source. The sites of phosphorylation are typically serine, threonine, or tyrosine residues (Table 1.3 shows the corresponding structures). Aspartic acid, histidine, and lysine may also be phosphorylated. Sulfation of tyrosine residues is another post-translational modification; however, it is not commonly observed. Disulfide bond formation, as in cyclic peptides, can also be considered a post-translational modification.

Hydroxylation is an important process in the stabilization of the structure of collagen. This oxidation process occurs for proline and lysine residues in certain sequences, to produce hydroxyproline (Hyp) and hydroxylysine (Hylys), shown in Table 1.3. The reaction is catalysed by prolyl 4-hydroxylase or lysyl 5-hydroxylase, respectively, in the presence of α-ketoglutarate and ascorbate (vitamin C).

In some calcium-binding proteins/peptides, Glu is post-translationally modified to γ-carboxyglutamic acid (Gla) (Table 1.3) via γ-glutamyl carboxylase. This residue in the dicarboxylate form is able to chelate Ca^{2+} ions.

The following sections include brief discussions of the post-translational modifications lipidation and glycosylation, as well as synthetic methods to achieve these modifications.

2.9 LIPIDATION

Lipidation occurs post-translationally by enzymatic attachment of lipid chains, in particular prenylation and acylation. Prenylation refers to the attachment of polyisoprenoids. These are often attached through a cysteine residue (Figure 2.22). Acylation is achieved naturally via thioester or amide bond formation. Some natural lipopeptides contain alkyl chains attached at the ε-amino group of lysine.

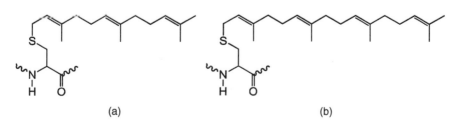

(a)   (b)

Figure 2.22 Peptide modified at cysteine residue with (a) farnesyl group, (b) geranylgeranyl group.

In preparing peptide–lipid conjugates, attachment of octanoyl, decyl, dodecyl, tetradecyl, hexadecyl, or octadecyl chains is common. These improve the pharmacokinetic properties (Section 5.2). These chains are often given non-systematic names based on the corresponding acids: caprylic (octanoic), capric (decanoic), lauric (dodecanoic), myristic (tetradecyl), palmitic (hexadecyl), or stearic (octadecyl).

N-terminally modified peptides are not recognized by aminopeptidases (N-terminus exopeptidases) and thus have longer circulation times *in vivo*. N-terminal alkylation can be performed using SPPS methods (Figure 2.1) with alkanoic acid coupling with HBTU in DIPEA. Palmitoylation, referring to attachment of C_{16} (hexadecyl) lipid chains, is a particularly common procedure since palmitoyl lipopeptides are expected to have good compatibility with lipid membranes, and they also show interesting self-assembly properties (discussed further in Section 3.11). Figure 2.23 shows amide-linked palmitoylated lipopeptides with palmitoyl lipidation at the N-terminus or the C-terminus. Lipidated peptides can exhibit improved oral bioavailability due to improved permeation across lipid membranes. This can also be achieved by glycosylation discussed in Section 2.10.

C-terminally lipidated peptides can be prepared via aminolysis of resin-bound ester (as in C-terminal palmitoylation, Figure 2.23b) or using amine-modified linkers, via the reactions shown in Figure 2.24. In a few cases, peptides can be linked via the phosphoryl head group of lipids. One example is the anticancer compound mifamurtide, discussed in Section 5.6.

Lipid chains can be attached to the ε-amine of lysine groups by deprotection and coupling as described above. A γ-L-glutamyl linker may be introduced to further increase (on top of the elimination of the ε-amine $-NH_3^+$ group) the negative charge on a peptide, which can improve binding to albumins, benefitting circulation times. In fact, attachment of lipid chains through the ε-amine group is used in the important therapeutic lipopeptides insulin detemir and liraglutide, both used to treat diabetes. These compounds are discussed further in Sections 5.3.3 and 5.3.5.

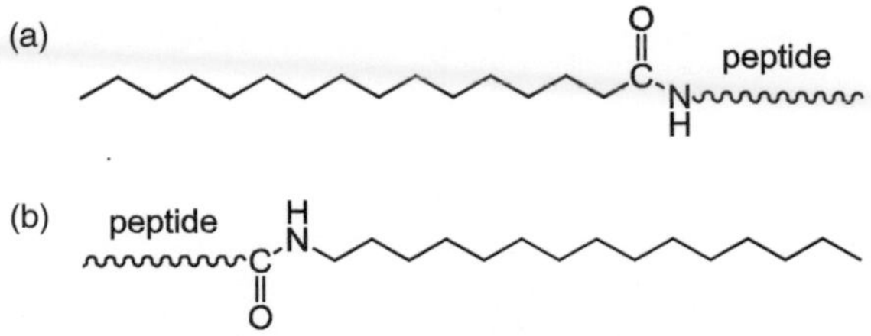

Figure 2.23 Lipopeptides prepared by (a) N-terminal palmitoylation, (b) C-terminal palmitoylation.

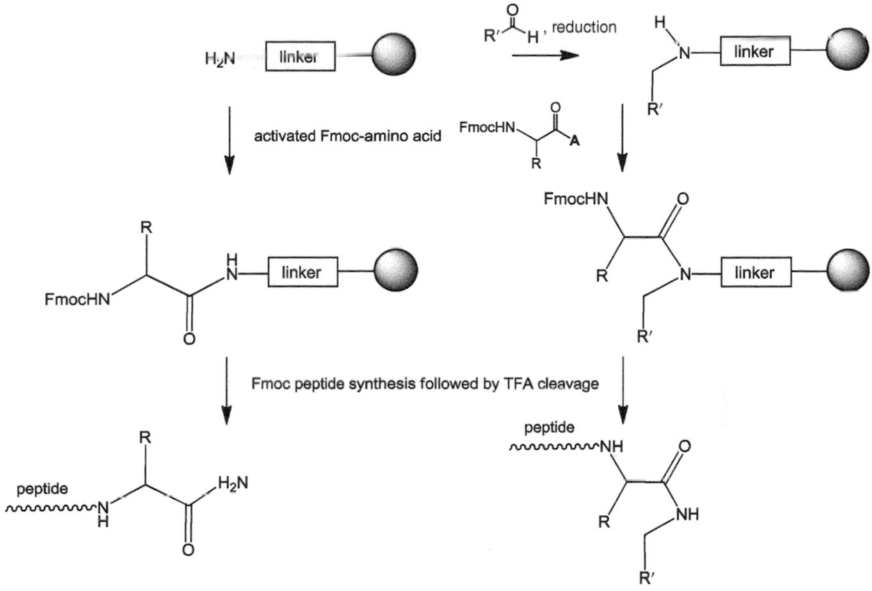

Figure 2.24 Synthesis of amidated peptides using amine-modified linkers.

2.10 GLYCOSYLATION

Glycosylation is an important post-translational modification of many proteins and of some peptides. It is achieved via glycosyltransferase enzymes. The saccharide units in the glycopolymer are attached to a peptide via N-glycosidic bonds at asparagine residues or via O-glycosidic bonds at serine or threonine residues (Figure 2.25).

The synthesis of glycopeptides may be done via coupling of a glycopolymer to a peptide on resin or using commercially available glycosylated residues. The former requires a linker that is stable to both acid and base conditions, as well as use of an activated sugar due to the low reactivity of the hydroxyl groups in saccharides. In addition, for N-glycosylated peptides, there is the problem of aspartimide formation (Section 2.2.6) when coupling glycosylamine and aspartic acid. This can be avoided using an N-protected aspartic acid α-benzyl ester. Therefore, the most common approach to synthesize glycopeptides uses glycosylated amino acids. Due to the hydrolysis of hydroxyl groups under acidic conditions mentioned above, these groups are usually reversibly protected using benzyl ether, acyl, or acetyl protecting groups. With the commonly used acetyl blocking groups, deprotection is achieved using sodium methoxide in methanol, hydrazine hydrate, or saturated ammonia in methanol.

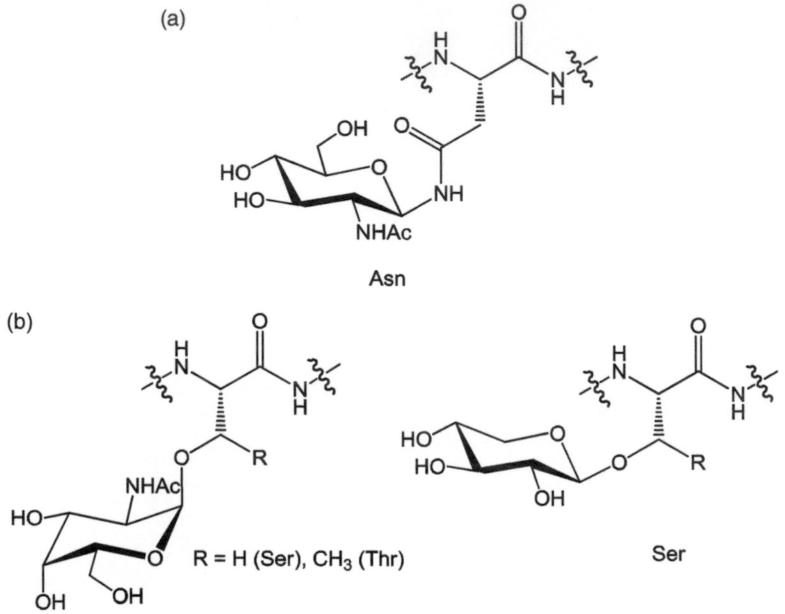

(a)

Asn

(b)

R = H (Ser), CH₃ (Thr)

Ser

Figure 2.25 (a) N-glycosylation of an N residue, (b) O-glycosylation of S or T residues.

Enzymatic glycosylation can in principle be performed on resin in SPPS, although this method has not yet been widely exploited.

2.11 POLYPEPTIDE POLYMERS AND CONJUGATES OF PEPTIDES AND POLYMERS

2.11.1 Polypeptides via NCA Polymerization

Well-defined polypeptide polymers may be prepared using living ring-opening polymerization (ROP) of N-carboxyanhydrides (NCAs) (Figure 2.26a). The NCA monomer is prepared via the Leuchs or Fuchs–Farthing approaches. The former involves the reaction of N-alkyloxycarbonyl amino acids with halogenating compounds, while the latter employs the use of phosgene (or a stable substitute) as the carbonyl source in the direct addition to a α-amino acid. The NCA-ROP reaction mechanism is shown in Figure 2.26b. It can be initiated using transition metal catalysts among other reagents. This approach can also be used to prepare block copolypeptides via sequential polymerization. High

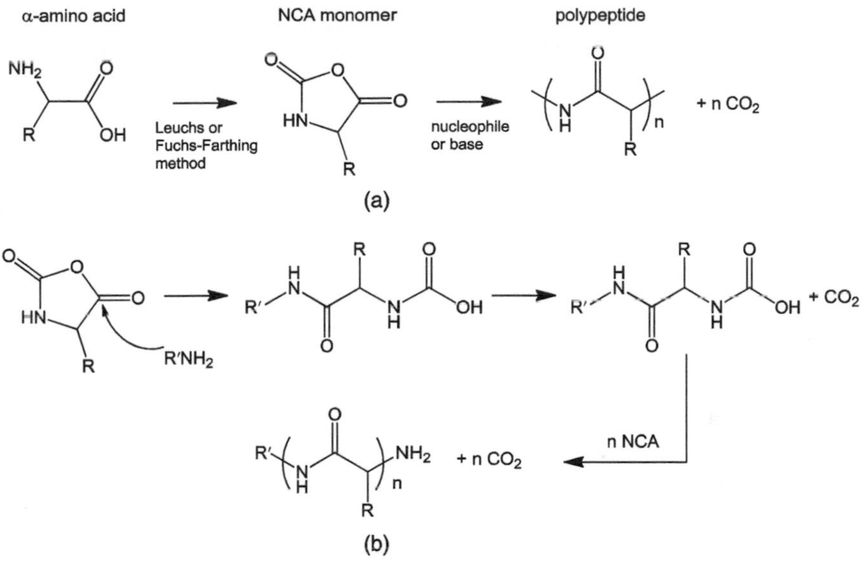

Figure 2.26 Synthesis of polypeptides via *N*-carboxyanhydride polymerization: (a) schematic reaction sequence, (b) mechanism of NCA polymerization initiated using nucleophilic amines

vacuum techniques can be employed in order to further improve the control of molecular weight and dispersity.

ROP methods provide multi-valent display of multiple peptide groups distributed along the polymer chain. High densities of peptide on brushes can be achieved using ROP and other methods such as reversible addition–fragmentation chain-transfer (RAFT) polymerization. RAFT and ATRP (atom transfer radical polymerization) are modern living free-radical polymerization methods, used to produce well-defined, low dispersity polymers. Multivalent display can be enhanced by attachment of peptides to star or branched polymer chains.

2.11.2 Coupling Chemistries

Bioconjugation is of considerable interest in the modification of peptide and protein structures and biological activity. Bioorthogonal reactions involve non-native, non-perturbing chemical moieties that can modify biomolecules (here: peptides and proteins) via selective reactions with ligands such as polymers, fluorescent probes, other biomolecules, etc. Native chemical ligation (NCL) and Staudinger ligation are examples of bioconjugation reactions. NCL was discussed in Section 2.4.2. The traceless Staudinger ligation reaction is shown in Figure 2.27f.

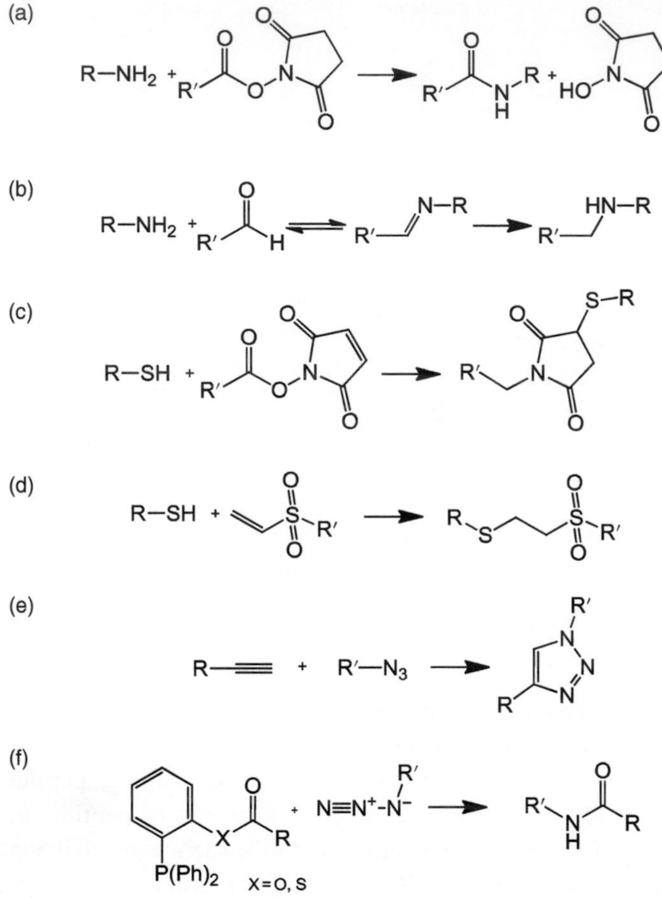

(a)

(b)

(c)

(d)

(e)

(f)

Figure 2.27 Representative coupling chemistries. In general R denotes a peptide chain and R′ a PEG chain, although in some cases these are interchangeable (e.g. reaction e). (a) An amine reacts with an NHS ester to form an amide bond. (b) An amine reacts with an aldehyde to produce a Schiff base, which can be reduced by borohydrides to produce a secondary amine linkage. (c) A thiol (in a cysteine residue) reacts with a maleimide derivative to produce a thioether bond, (d) A thiol-containing peptide reacts with a vinylsulfone-modified PEG to produce a thioether bond. (e) The Cu-catalysed alkyne-azide 'click reaction' produces a triazole. (f) Traceless Staudinger ligation.

Attachment of polyethylene glycol (PEG), known as PEGylation of peptides, is of great interest since PEG provides 'steric stabilization', enhancing *in vivo* circulation as it provides a barrier around the peptide-reducing degradation. PEG is a readily available polymer which is synthesized in a wide range of molar masses (using a variety of methods) with well-defined properties and good *in vivo* stability. A number of PEGylated therapeutic molecules have reached the market.

Several strategies are available to synthesize PEG-peptide conjugates. Many of these can also be used to create other polymer–peptide conjugates. Methods can be summarized as: (i) coupling techniques, (ii) peptide synthesis from PEG chains, so-called 'grafting to' chemistry, (iii) PEG synthesis from peptide, so-called 'grafting from', and (iv) attachment of PEG to side chains. These methods are discussed in turn in the following sections.

A wide range of chemistries are available to couple polymers and peptides, exemplified in the following by using PEG as a commonly coupled polymer. Examples of some common coupling chemistries are shown in Figure 2.27. This is not an exhaustive list of methods used to link PEG polymers and peptides. Amine groups at the N-terminus or on lysine side chains are common sites for PEG attachment using hydroxyl- or aldehyde-functionalized PEG. A widely used method in current use involves *N*-hydroxysuccinimide (NHS) esters, which are highly reactive towards amines at physiological pH (Figure 2.27a). Potential hydrolysis of the ester bond of succinylated PEG can be avoided by use of a linking carbonate group.

Rather than preparing polymers with reactive end groups, *in situ* activation using carbodiimides is possible, and this has been used to couple the carboxylic group of succinylated PEG to various peptides. Aldehyde-terminated polymers may be reacted with N-terminal amines (Figure 2.27b).

Cysteines are an attractive target for conjugation of polymers to peptides. A number of thiol-reactive groups can be used to couple end-functionalized polymers to cysteine-terminated peptides (e.g. Figure 2.27c,d). One widely used chemistry employs maleimides that react selectively with the thiols of cysteine residues by Michael addition in the pH range 6.5–7.5 (Figure 2.27c). To prepare ATRP initiators containing maleimides, protection is required. Amine-terminated PEG has been converted to a maleimide by reaction with maleic anhydride. Maleimide-modified RAFT chain transfer agents (CTAs) have also been employed. Another chemistry that can be used to link polymers to cysteine residues employs vinyl sulfone-terminated polymers (Figure 2.27d). Thiol-ene chemistry may also be employed to link alkene-terminated polymers and cysteine-containing peptides.

The well-known cycloaddition 'click' reaction between alkynes and azides (Figure 2.27e) has been widely used in the synthesis of bioconjugates, including polymer–peptide conjugates. By using click reactions, grafting efficiencies approaching 100% are possible.

An example of a traceless Staudinger ligation is shown in Figure 2.27f. Other phosphino(thio) esters may be used. The reaction avoids the necessity for an N-terminal cysteine residue in the NCL reaction. The phosphonium species is liberated in the final hydrolysis step. The traceless reaction avoids the formation of phosphine oxide in the original Staudinger ligation reaction.

On-resin coupling of PEG-CH$_2$-COOH to N-terminal peptide resins can be performed via acylation.

2.11.3   Peptide Synthesis from Polymer Chains ('Grafting to')

N-Carboxyanhydride (NCA) polymerization from PEG macroinitiators enables the synthesis of a range of polymer–peptide conjugates. NCA polymerization of homopolypeptides such as poly(L-proline) as well as copolymers from PEG macroinitiators has been demonstrated. A variety of chemistries have been employed using PEG as a cleavable support for peptide synthesis (prior to cleavage PEG/peptide conjugates are produced). Some PEGylated resins are available commercially.

PEG chains have been used as supports for liquid-phase peptide synthesis, via direct esterification. In general, the synthesis is carried out in organic solvents, although some examples in water have been reported, using EDC (1-ethyl-3-[3-(dimethylamino)propyl] carbodiimide) instead of DCC to activate the ester.

2.11.4   Polymer Synthesis from Peptide ('Grafting from')

In the grafting from method, initiators for polymerization methods such as ATRP may be incorporated at the peptide termini. This method has been widely used to prepare protein–polymer conjugates, as well as polymer–peptide conjugates. A RAFT agent can also be appended to a solid-phase synthesized peptide and used as a macroinitiator for RAFT.

2.11.5   Synthesis of Side Chain Polymer/Peptide Conjugates

Polymers (and other moieties) may be linked to peptides as side chains via functionalizable groups. For lysine, reactions such as those in Figure 2.27a,b can be used, while for cysteine, reactions such as those in Figure 2.27c,d are suitable. Figure 2.28 shows some further common side group reactions. The reaction in Figure 2.28a involves an electrophilic attack by an isocyanate/thioisocyanate, leading to stable (thio)urea linkages. The reaction in Figure 2.28b corresponds to amidation with imido esters. This

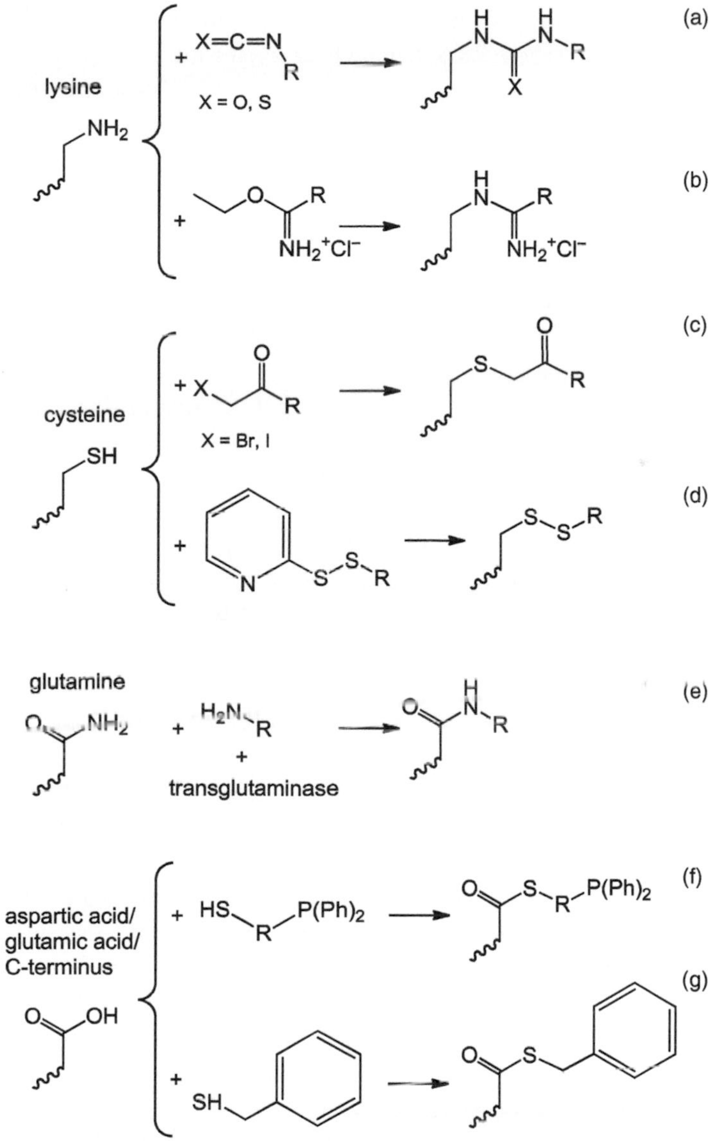

Figure 2.28 Side group reactions for peptide functionalization. For lysine and cysteine, additional chemistries are shown in Figure 2.27a–d.

reaction maintains the charge on the residue. The reaction in Figure 2.28c leads to the formation of stable thioether bonds from α-halocarbonyl compounds. These bonds can also be formed using maleimides (Figure 2.27c) or vinyl sulfones. In Figure 2.28d, a (reversible) disulfide bridge is formed using an orthopyridyl disulfide (a similar reaction occurs with methanethiosulfonates). Glutamine can be transamidated using the enzyme transglutaminase (Figure 2.28e), which catalyses an acyl transfer reaction between the γ-carboxamide group of glutamine residues and amino groups on the polymer ligand. Carboxylic acid groups in Asp, Glu, E, and the C-terminus can be used to couple polymers via carbodiimide coupling (and this can be performed using SPPS protocols). Phosphino thiols may also be used to form thioesters suitable for Staudinger ligation (Section 2.11.2) as shown in Figure 2.28f. The thioesters generated in the reaction shown in Figure 2.28g are useful in NCL (Section 2.4.2). Some reactions are also available for coupling through aromatic residues. A variety of chemistries are available for coupling via the N-terminus. The pK_a of the N-terminus is different to that of lysine and arginine side chains (Table 1.2) and this enables selective N-terminal modification by methods used for lysines as discussed above. Residues adjacent to the N-terminus may also be used in selective coupling reactions.

Also, peptides may be incorporated as side chains in polymers via incorporation in the monomer or post-polymerization modification (using for example, appended coupling groups such as click motifs).

2.12   NON-RIBOSOMAL PEPTIDE SYNTHESIS

NRPS is used by nature to produce several complex peptides and peptide derivatives. The ribosome is a large complex of RNA and proteins that synthesizes proteins in cells. Non-ribosomal synthesis occurs via enzymes which can couple specific residues, for example the tripeptide glutathione (L-γ-glutamyl-L-cysteinylglycine, H-isoGlu-Cys-Gly-OH) (Section 5.4) is synthesized via NRPS via two enzymes. The first, γ-glutamyl synthetase, couples glutamic acid and cysteine through the γ-carboxyl group of the glutamic acid. The second enzyme, glutathione synthetase, adds the glycine residue via a peptide bond. Many peptide antibiotics are synthesized through NRPS via large multifunctional enzymes. Post-translational modifications are also often made non-ribosomally.

2.13   PURIFICATION AND ANALYSIS METHODS

2.13.1   Mass Spectrometry

Mass spectrometry is used to check the molar mass of synthesized peptides. The method actually measures the mass-over-charge ratio (m/z) of the molecules. The method usually used for lower molar mass peptides is electrospray ionization mass spectrometry (ESI-MS). For higher molar mass peptides or peptide conjugates where electrospraying does not work, other methods, especially MALDI-MS (matrix assisted laser desorption/ionization mass spectrometry) may be employed. The 'ionization' in electrospray ionization is a soft ionization method in which a solubilized sample is passed through a needle at high voltage. Ionization processes occur under high vacuum in the mass spectrometer. In MALDI, the ionization occurs by laser dissociation of sample which is co-deposited with an organic matrix compound that often has a conjugated aromatic ring structure. The mass detection analysis is usually done via quadrupoles, ion traps, or time-of-flight analysers.

It has to be considered that elements present in peptides, especially carbon, are present as mixed isotopes (e.g. ^{12}C and ^{13}C are present at naturally at 98.9% and 1.1% abundances, respectively) and in high resolution instruments the monoisotopic mass can be distinguished from the average mass. The exact mass of an isotopic species is obtained by summing the masses of the individual isotopes of the molecule.

Peptides can also be sequenced by tandem MS (MS/MS) in which two mass spectrometers are coupled. The first separates ions according to their mass/charge ratio. These are selected and further fragmented (by collision-induced dissociation or ion-molecule reaction or photodissociation) for analysis in the second mass spectrometer which detects the ions according to their mass/charge ratio. Because of the complexity of peptide fragmentation, a notation system was introduced to describe the fragmentation pattern. This is shown in Figure 2.29. The N-terminal charged fragment ions are classed as a, b, or c, while the C-terminal charged ones are classed as x, y, or z. Immonium ions are labelled a or x. When the peptide bond is fragmented, b ions are those in which the amino terminus retains the charge, in contrast to y ions where the carboxy terminus of the fragment retains the charge. When the $N–C_\alpha$ bond is cleaved, the corresponding ions are labelled c and z.

More advanced mass spectroscopy methods, measuring ion mobility, can be used to probe the formation of oligomeric peptide structures such as those formed by amyloid peptides.

2.13.2   HPLC and Other Chromatographic Methods

Peptide purity is commonly analysed by reverse-phase high performance liquid chromatography (RP-HPLC) using a silica C_8 or C_{18} bead column

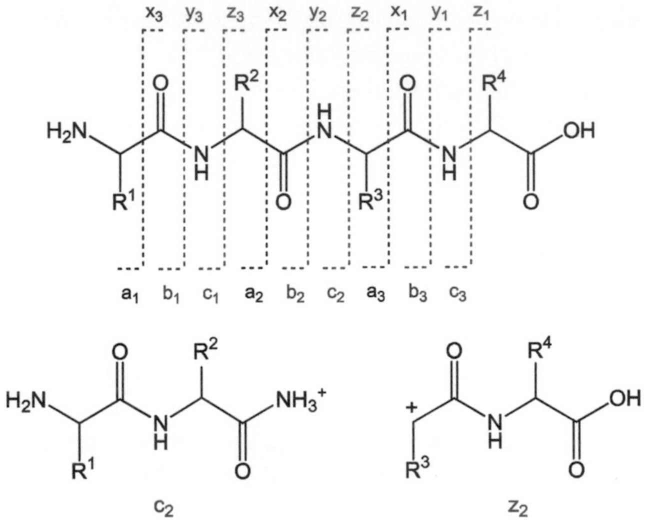

Figure 2.29 Peptide ion fragmentation labelling system for tandem MS along with two representative ion fragments.

(3–5 μm). This is a high resolution method which involves separation via partitioning of molecules between a mobile phase and a hydrophobic stationary phase. The eluent is usually an acetonitrile/water gradient with trace TFA (often 0.1%) to improve resolution (reducing interactions between silanol groups and basic residues). The signal is detected by UV absorbance at 210–230 nm. The area under the main peak due to peptide product relative to any side peaks provides a commonly quoted measure of peptide purity.

HPLC may also be used in a semi-preparative fashion for separation of peptides during synthesis. This can be done manually or using automated HPLC systems that include autosamplers.

Ion-exchange chromatography (IEC) can be used to separate peptides based on net charge. An anion-exchange column is functionalized with positively charged groups, whereas a cation-exchange column will contain negatively charged groups. The pH of the starting buffer must be carefully controlled since it determines the charges on the peptides to be separated. The starting buffer pH should be at least one pH unit above the pI of the peptide to ensure binding to the matrix. To elute the peptide of interest, a solution with a higher concentration of counterion is added to the column. The affinity of a peptide for the column can also be changed via the pH of the column buffer. In addition, IEC is used to exchange the counterion in a peptide salt, for example replacing TFA with acetate.

Other chromatographic separation methods including capillary electrophoresis, affinity chromatography, or gel filtration chromatography are occasionally used. Capillary electrophoresis involves the separation of peptides based on their differential migration in an electric field. Affinity chromatography exploits the selective binding of targets to bound ligands. Gel filtration chromatography is also known as gel permeation chromatography (GPC) or size exclusion chromatography (SEC) and is generally used for larger macromolecules, sometimes including larger peptides.

2.13.3 Chemical Tests

The Kaiser test, which involves the reaction of ninhydrin with primary amines (Figure 2.30), is used to assess coupling during sequential peptide synthesis. The Schiff base chromophore product (2-(1,3-dioxoindan-2-yl)iminoindane-1,3-dione, Ruhemann's violet) has a characteristic purple colour.

Edman degradation can be used to sequence short peptides. It involves the removal of amino acids one at a time from the N-terminus using phenyl isocyanate (Figure 2.31). In the first step, the N-terminal amino group is reacted with phenylisocyanate to produce the phenylthiocarbamate (PITC) derivative. Excess PITC is removed with an anhydrous acid such as TFA to

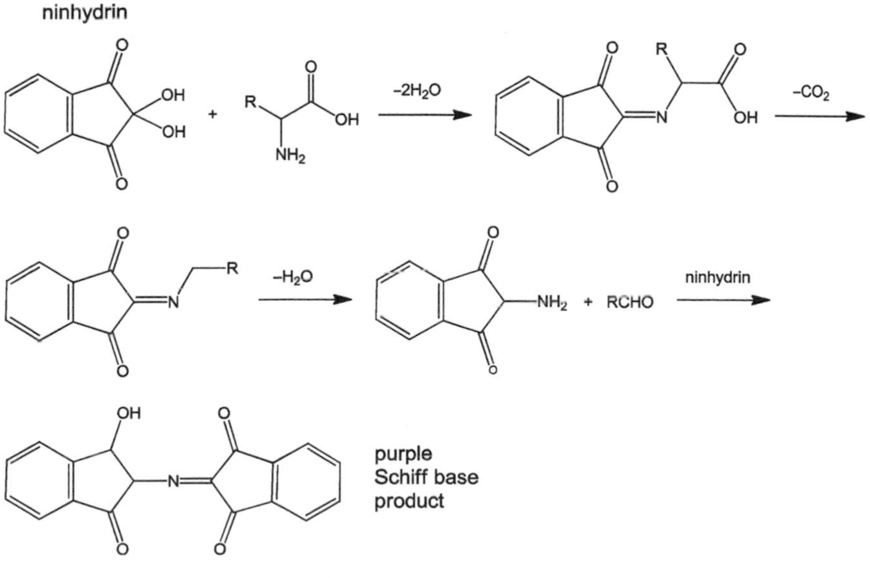

Figure 2.30 Reaction of ninhydrin with primary amines.

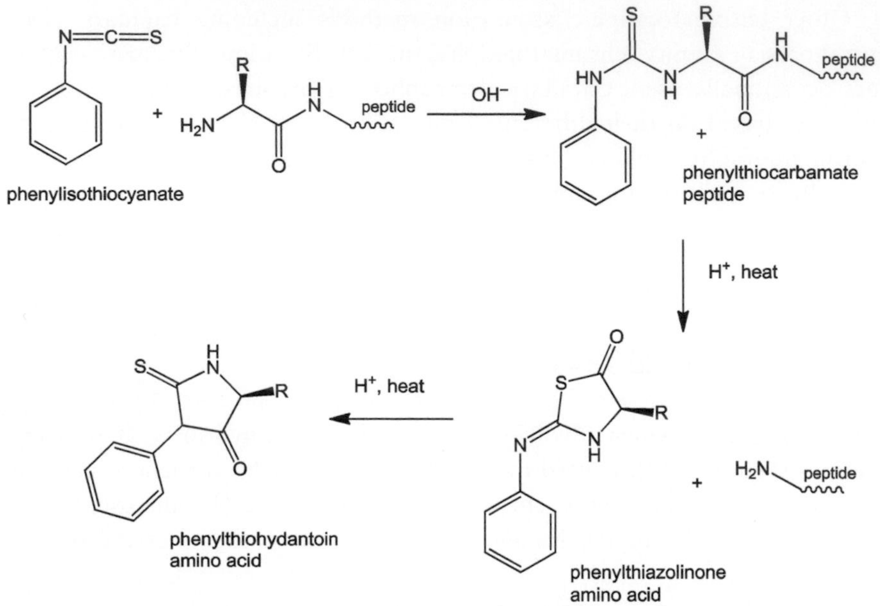

Figure 2.31 Scheme of the Edman degradation reaction.

produce a thiazolinone derivative. This is selectively extracted in organic solvent and then undergoes rearrangement in hot TFA to give a stable phenylthiohydantoin–amino acid derivative. This is analysed by RP-HPLC, TLC (thin-layer chromatography), or electrophoresis. The method is advantageous since it requires only low pico-molar quantities of sample.

BIBLIOGRAPHY

Canalle, L.A., Löwik, D.W.P.M., and van Hest, J.C.M. (2010). Polypeptide-polymer bio-conjugates. *Chemical Society Reviews* 39: 329–353.

Chan, W. and White, P. (1999). *Fmoc Solid Phase Peptide Synthesis: A Practical Approach.* Oxford: Oxford University Press.

Cheng, J. and Deming, T.J. (2012). Synthesis of polypeptides by ring-opening polymerization of α-amino acid N-carboxyanhydrides. In: *Peptide-Based Materials* (ed. T.J. Deming), 1–26. Berlin: Springer-Verlag.

Doonan, S. (2002). *Peptides and Proteins.* Cambridge: RSC.

Gauthier, M.A. and Klok, H.A. (2008). Peptide/protein-polymer conjugates: synthetic strategies and design concepts. *Chemical Communications*: 2591–2611.

Hamley, I.W. (2014). PEG-peptide conjugates. *Biomacromolecules* 15: 1543–1559.

Hermanson, G.T. (2008). *Bioconjugate Techniques.* New York: Academic Press.

Jensen, K., Tofteng Shelton, P., and Pedersen, S.L. (2013). *Peptide Synthesis and Applications*, 2e. New York: Humana Press.

Jones, J. (2002). *Amino Acid and Peptide Synthesis*. Oxford University Press.

Nielsen, P. (2004). *Pseudo-Peptides in Drug Discovery*. Weinheim, Germany: Wiley-VCH.

Palomo, J.M. (2014). Solid-phase peptide synthesis: an overview focused on the preparation of biologically relevant peptides. *RSC Advances* 4: 32658–32672.

Sewald, N. and Jakubke, H.-D. (2002). *Peptides: Chemistry and Biology*. Weinheim: Wiley-VCH.

Sunna, A., Care, A., and Bergquist, P.L. (2017). *Peptides and Peptide-based Biomaterials and their Biomedical Applications*, Advances in Experimental Medicine and Biology. Cham, Switzerland: Springer.

Tulla-Puche, J., El-Faham, A., Galanis, A.S. et al. (2015). Methods for the peptide synthesis and analysis. In: *Peptide Chemistry and Drug Design* (ed. B.M. Dunn), 11–73. New York: Wiley.

Wibowo, S.H., Sulistio, A., Wong, E.H.H. et al. (2014). Polypeptide films via N-carboxyanhydride ring-opening polymerization (NCA-ROP): past, present and future. *Chemical Communications* 50: 4971–4988.

3

Amyloid and Other Peptide Aggregate Structures

3.1 INTRODUCTION

This chapter discusses the structures of aggregated peptides and the methods to characterize them. Amyloid-forming peptides are considered along with other assemblies, including structures such as peptide nanotubes, micelles, and vesicles formed by peptide conjugates including lipidated peptides and polymer–peptide hybrids. In addition, peptide hydrogels are covered in this chapter, including a discussion of their use in tissue engineering among other applications.

The structures considered in this chapter have the common feature that they extend beyond the secondary structure. Although amyloid is a natural structure based on β-sheets, the fibrils are highly extended, beyond their range in folded protein and peptide structures. Peptide hydrogels are structures formed when peptide fibrils form networks, often based on non-covalent cross-linking junction points.

The other peptide structures discussed in this chapter are formed through self-assembly of designed peptides or peptide conjugates. Nanotubes, micelles, and vesicles are rarely observed for natural peptides *in vivo*, although it is possible to induce some native peptides and lipopeptides to form these nanostructures under defined conditions *in vitro*.

Understanding the formation of amyloid is the subject of great research activity because it characterizes many conditions, especially neurodegenerative diseases including Alzheimer's disease (AD), Parkinson's disease, Huntington's disease, and amyotrophic lateral sclerosis (ALS) as well

Introduction to Peptide Science, First Edition. Ian W. Hamley.
© 2020 John Wiley & Sons Ltd. Published 2020 by John Wiley & Sons Ltd.

as type II diabetes. Table 3.1 presents a longer list of amyloid diseases. So-called functional amyloid structures such as curli fibres are produced by bacteria and fungi as part of the production of extracellular matrix (ECM) during biofilm formation (which is also discussed in Chapter 4 on antimicrobial peptides). Most recently, amyloid fibrils have been engineered for applications in catalysis, water purification, carbon dioxide sequestration, and filtration since they can form high density meshes with high surface area. They also have potential as high performance fibres since the Young's modulus (which is a measure of stiffness) of amyloid fibrils can be up to GPa (gigapascals) and the tensile strength of some amyloid materials such as silk is comparable to that of steel. Their potentially large anisotropy may

Table 3.1 Conditions associated with amyloid-forming peptides and proteins.

Condition	Peptide/Protein	Indication
Alzheimer's disease	Amyloid β peptide	Extracellular amyloid plaques
Amyloid light-chain (AL) amyloidosis	Immunoglobulin light chains	Amyloid deposits
Amyotrophic lateral sclerosis (ALS)	Superoxide dismutase I (SOD I)	Intraneuronal inclusions
Aortic medial amyloidosis	Medin	Amyloid deposits
Corneal amyloidosis associated with trichiasis	Lactoferrin	Amyloid deposits
Creutzfeldt–Jakob disease (CJD) and other spongiform encephalopathies	Prion protein	Extracellular plaques
Dementia with Lewy bodies	α-synuclein	Lewy bodies
Dialysis-related amyloidosis	β_2-microglobulin	Amyloid deposits in blood
Familial amyloidotic polyneuropathy	Transthyretin mutants	Amyloid deposits
Familial British dementia	ABri	Extracellular plaques
Familial Danish dementia	ADan	Extracellular plaques
Frontiotemporal dementia	tau	Neurofibrillary tangles
Huntington's disease	Huntingtin with polyQ-expansion	Intracellular inclusion bodies
Isolated atrial amyloidosis	Atrial natriuretic factor (ANF)	Amyloid deposits
Medullary thyroid cancer	Calcitonin	Amyloid deposits
Parkinson's disease	α-synuclein	Lewy bodies
Type II diabetes	Amylin, also known as islet amyloid polypeptide (IAPP)	Granule formation

also be exploited to template the deposition of arrays of inorganic materials such as metal nanoparticles or graphene sheets.

This chapter first discusses both natural amyloids and synthetic designed peptides that form β-sheets, which comprise the secondary structure element of amyloid. Amyloid peptides, both synthetic and nature-equivalent, e.g. the Amyloid β peptides Aβ40 and Aβ42, with 40 or 42 residues respectively (Section 3.3), may be prepared synthetically. However, other amyloids are proteins or long polypeptides that are not amenable to chemical synthesis methods. Here, and indeed also for large-scale production of shorter peptides, genetic engineering methods may be employed. Yeast models are of particular interest. A number of simplified models for amyloid systems, including amyloid-expressing fruit flies, have been developed that have a fast life cycle so that amyloid-related neurodegenerative effects can be studied over a realistic timescale.

This chapter is organized as follows. Section 3.2 discusses some general common features of amyloid. This is then followed by a more detailed discussion in Section 3.3 of the Amyloid β peptides, which are intensely studied amyloid peptides, implicated in the progression of AD. Then the mechanisms and kinetics of amyloid propagation are analysed in Section 3.4. The toxicity of amyloid oligomers and their connection to amyloid diseases is considered next in Section 3.5. The fibrillization of smaller amyloid peptides, often used as model systems, is discussed in Section 3.6. Section 3.7 provides a summary of functional amyloid and bioengineering applications of amyloid materials. While most peptide fibrils are based on β-sheet structures, it is possible to design coiled coil peptides to associate into fibrils. This is briefly outlined in Section 3.8. Peptide hydrogels are useful materials that result when peptide fibrils form cross-linked networks, especially via non-covalent self-assembly. The formation of peptide gels and their applications in cell culture and drug delivery are discussed in Section 3.9.

The remainder of the chapter concerns other non-fibrillar peptide aggregates. Section 3.10 provides an introduction to peptide nanotube structures, including a discussion of distinct forms and mechanisms of assembly along with examples of peptides, both natural and synthetic, which can form such structures.

Conjugation of a peptide to lipids or polymers can lead to distinct aggregate nanostructures, self-assembly being driven by amphiphilicity due to attachment of a hydrophilic peptide to a hydrophobic lipid or a hydrophilic polymer to a hydrophobic peptide, for example. Self-assembly can also result from the architecture of the peptide, for instance surfactant-like, bolaamphiphilic, and ionic-complementary peptides can all aggregate into ordered nanostructures. These topics are discussed further in Section 3.11.

The chapter concludes with Section 3.12, which discusses the main methods used to characterize peptide aggregate structures.

3.2 AMYLOID

The term 'amyloid' refers to protein deposits resembling those first observed for starch, as 'amyloid' originally meant 'starch-like'. It is now specifically associated with proteins and peptides forming fibrils based on the cross-β structure in which the peptide backbone is orthogonal to the fibril axis (Figure 3.1). This pattern is characterized by X-ray diffraction on aligned fibres, as shown schematically in Figure 3.1. The β-sheets self-assemble into fibres, which may comprise a structure such as a bundle of twisted β-sheets. This is shown in Figure 3.1, which is a real structure of an amyloid fibre based on solid state NMR (nuclear magnetic resonance) spectroscopy. Twisting of fibrils results from the preferred right-handed twist of β-sheets (see Section 1.4). Both parallel and antiparallel arrangements of β-sheets have been observed in amyloid fibrils. Figure 3.2a shows a typical electron micrograph image of amyloid fibrils from Aβ42 (Section 3.3), with regions of twisting apparent.

The signature in an X-ray scattering pattern of the cross-β structure is a 4.7–4.8 Å meridional reflection, corresponding to the spacing between peptide backbones, and an equatorial reflection at 8–12 Å that is broader and which corresponds to the stacking periodicity of β-sheets. The range of values reflects different side chain dimensions; even smaller values are

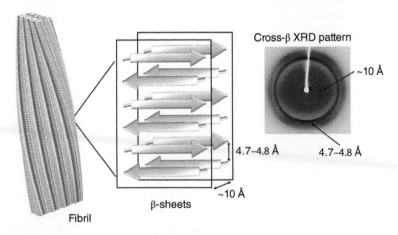

Figure 3.1 Hierarchical structure of amyloid fibrils and 'cross-β' X-ray diffraction pattern.

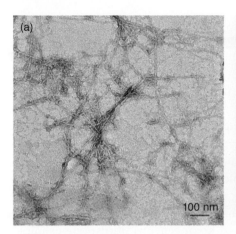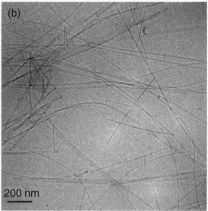

100 nm   200 nm

Figure 3.2 TEM images of amyloid fibrils: (a) negative stain TEM image from a solution of Aβ42, (b) cryo-TEM image from a mixed solution of peptide C_{16}-KKFF and C_{16}-RGD.

possible for peptides with long sequences of residues with small side chains such as polyalanine repeats. These X-ray diffraction patterns are obtained from fibre diffraction experiments, using aligned stalks (see Section 1.5.5). These measurements are performed since most amyloid peptides do not crystallize, almost by definition since the amyloid structure is a highly thermodynamically favourable non-crystalline structure. Only a few short fragment amyloid peptides have been crystallized such that single-crystal X-ray diffraction structures could be obtained. These form structures called steric zippers because the side chains are packed in a zippered arrangement. For most amyloid peptides, structures are obtained from fibre diffraction along with NMR. An example of a solution NMR structure for Aβ42 is shown in Figure 3.1.

Electron microscopy, especially transmission electron microscopy (TEM) and atomic force microscopy (AFM) (both discussed further in Section 3.12.3), are used to image amyloid fibrils. Methods to obtain high-resolution TEM (and cryogenic-TEM) images have developed rapidly and these techniques have been used to obtain detailed information on the structure of fibrils of many important amyloid peptides such as Aβ. Image reconstruction methods and/or sample tilting procedures can now be used to obtain 3D images of amyloid fibrils at a resolution down to about 3 Å.

There is typically considerable variation in the structure of amyloid fibrils, which is termed polymorphism. This can involve the presence of multiple forms of fibrils with different degrees of twist or number of proto-filaments comprising the fibrils (leading to differences in fibril width) or their packing arrangement, or variability in the length of fibrils. Polymorphism can arise among the fibrils within a given sample (intra-sample polymorphism) or from variation between samples (inter-sample variability). Figure 3.2b shows an example of a cryo-TEM image showing intra-sample polymorphism, with variable twisting and number of filament subunits within fibrils visible. Inter-sample polymorphism can result from changes in sample conditions, such as use of different pH or buffer, or changes in sample preparation conditions (stirring or agitation can break fibrils for example), or changes in incubation time or sample preparation temperature or concentration. Seeding of amyloid samples also leads to distinct polymorphs. In some cases, different polymorphs can be propagated by seeding, as with different strains of the amyloid form of the prion protein, PrP^Sc. Some amyloid systems can form well-defined homogeneous fibrils; one example is the paired helical filaments (PHFs) formed as aggregates of the tau protein. Careful preparation methods can also be used to isolate particular polymorphs of Aβ and other amyloid peptides. Designed peptides can also assemble into relatively homogeneous amyloid fibril structures.

The formation of fibrils is symptomatic of many amyloid diseases such as AD and Creutzfeldt–Jakob disease (CJD). Other amyloid diseases are listed in Table 3.1. The entries in this table show that amyloid diseases affect many organs. More extensive lists of amyloid-related diseases, of which there are a considerable number, can be found in the literature. The amyloid is often present as aggregates called plaques or in Lewy bodies, which are clumps of fibrils with a dense core. As discussed in Section 3.5.1, it is now believed that the toxic species may not be the fibrils, but rather oligomeric forms formed on the pathway to fibrillization (Figure 3.3). The formation of amyloids by peptides is much simpler than that of proteins since complex unfolding pathways of proteins do not have to be considered for peptides, although there

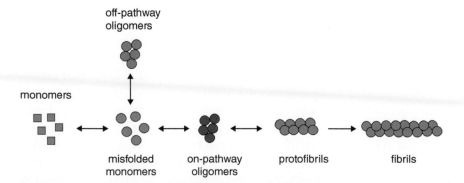

Figure 3.3 Oligomerization and fibrillization pathways. Species in green are non-toxic, those in red are toxic, and those in blue may be toxic or not.

may be a conformational transition if the initial peptide conformation is not β-sheet (these states are termed misfolded monomers in Figure 3.3).

Many peptides form amyloid and it is hard to give general rules about which will do so. There are natural peptides, including those responsible for diseases (Table 3.1), and fragments of these peptides also form amyloid (some are discussed in Section 3.6). In addition, many peptides have been designed that form amyloid structures by designing sequences that favour inter-residue hydrogen bonding and side chain packing. Peptides can be assessed for aggregation propensity based on the content of particular residues, as provided in several scales such as the Chou–Fasman scale (see Figure 1.9), or by using online resources that are built from protein databases or which use physicochemical parameters to predict aggregation tendency. Examples such as TANGO, Waltz, Camsol, and S2D are listed in Table 1.6.

3.3   AMYLOID β

3.3.1   Relation to Alzheimer's Disease and Features of Aβ Peptides

Alzheimer's disease (AD) is named after Alois Alzheimer who first described the condition in 1906. He observed large aggregates (called plaques) in tissue from the cortex of a patient suffering from dementia. The aggregates were stained using Congo red (see Figure 4.21) and iodine. Later, the plaques were shown to be enriched in β-sheet fibrils.

AD is the most common cause of dementia (representing around 50–80% of all cases), with an estimated 18 million people worldwide currently affected by the condition (according to the World Health Organization). Its incidence increases dramatically with age, and the number of people with dementia is set to double in the next 20 years. The aggregation of Aβ peptides into oligomers or fibrils is now implicated as a key process associated with the progression of AD. The pathology of AD comprises amyloid plaques in brain tissue that are rich in Aβ, and neurofibrillary tangles in the hippocampus, amygdala, and association neocortex. The protein tau also has a role in AD progression, its processing occurring downstream of Aβ accumulation. Proteins including NAC (non-beta-amyloid component) are also co-deposited along with Aβ in plaques. NAC comprises residues 61–95 of α-synuclein, which is also involved in amyloidoses with Lewy bodies such as Parkinson's disease (Table 3.1). As well as deposition in plaques, levels of soluble Aβ in blood and cerebrospinal fluid (CSF) are elevated in AD sufferers who are showing indications of cognitive decline.

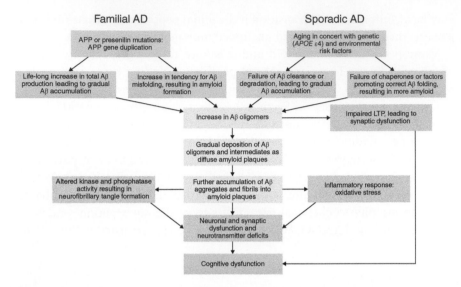

Figure 3.4   The amyloid hypothesis for AD. LTP (long-term potentiation) refers to synaptic signalling between neurons.

The processes underlying the Aβ hypothesis by which this peptide leads to AD are summarized in Figure 3.4. This is still a hypothesis since (among other reasons) it has not yet led to a successful therapy, and other mechanisms for this disease have been and are being proposed.

Alzheimer's disease (AD) accounts for ~70% of all late-onset dementia cases. Most cases occur relatively late in life, although ~5% occur in patients under 60 years old. These cases are termed early-onset or familial Alzheimer's disease (FAD). Genetic mutations associated with the Amyloid precursor protein (APP) or presenilin (both are transmembrane proteins) have been linked to these conditions. Sporadic or late-onset AD is associated with *APOE* gene variants (alleles). This gene encodes for a class of lipid-binding proteins called apoliproprotein E. Those carrying one or two copies of the *APOE* ε4 allele (sometimes termed ApoE4) are at higher risk for AD. In AD, neurodegeneration is estimated to start 10–30 years before clinical symptoms are detected. As discussed further in Section 3.5.1, the toxic form of Aβ is now believed to be oligomers rather than fibrils. The mode of activity is proposed to involve ion channel formation (especially calcium ion channels) in neuronal cell membranes. Oligomerization of Aβ occurs intracellularly. Extracellular Aβ may originate from intraneuronal sources, and a dynamic equilibrium may exist between these pools. Since Aβ is produced via cleavage of APP in membranes, its sites of production include the plasma membrane, but Aβ is also produced within the cell in the Golgi and endoplasmic reticulum, as well as endosomes and lysosomes.

The importance of lipid interactions with Aβ is highlighted by the fact that *APOE*, which is a key genetic risk factor for AD, is involved in lipid metabolism. Lipid membranes have a number of important roles in modulating amyloid fibrillization. These include: (partially) unfolding the peptide, increasing the local concentration of peptide bound to the membrane, orienting the bound peptide in an aggregation-prone manner, and varying the penetration depth into the membrane affecting the nucleation propensity. Lipid rafts (phase-separated lipid membrane domains, enriched in particular lipids) are implicated in Aβ dimer and oligomer formation, and may provide platforms for selective deposition of different Aβ aggregates. Aβ, which is a cationic peptide at neutral pH, interacts with anionic lipid membranes through electrostatic interactions, depending on pH. On the other hand, it has been reported that Aβ can interact with cationic or zwitterionic lipids as readily as anionic lipids. This suggests that association of Aβ with lipid membranes may be driven to a substantial extent by hydrophobic interactions with hydrophobic regions in the peptide.

Oxidative stress may play an important role in the age-dependent susceptibility to AD. Oxidative stress involves the production of free radicals (especially hydroxyl radicals) in the presence of metal ions, and this can influence metabolism and also promote Aβ aggregation. The free radicals can cause increased lipid peroxidation, and the formation of associated by-products, as well as protein and DNA oxidation in the AD brain. Diminished mitochondrial energy metabolism may play a role in AD pathogenesis, due at least in part to reduced cyclooxygenase (COX) activity. Excitotoxicity may also play a role in neuronal cell death. Excitotoxicity is the overstimulation of N-methyl-D-aspartate (NMDA) or 2-amino-3-(5-methyl-3-oxo-1,2-oxazol-4-yl)propanoic acid (AMPA) receptors (NMDARs and AMPARs respectively) by glutamate or aspartate, leading to neuronal hyperexcitability and death. Excitotoxicity can also generate excess reactive oxygen species (ROS). The role of oxidative stress is evidenced by the presence of protein glycation end products in Aβ aggregates, as well as an increase in the number of activated microglial cells. The inflammation that results from oxidative stress as well as Aβ- (and tau-) induced neurodegeneration has an important role in AD pathology.

Figure 3.5 shows the sequence of two variants of the Amyloid peptide found in humans – Aβ40 and Aβ42 (this notation will be used consistently for the whole peptide with the number of residues indicated). These Amyloid β peptides are produced by proteolytic cleavage of the APP, a transmembrane protein of unknown function. Part of the sequence of APP is also shown in

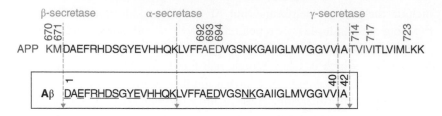

Figure 3.5 (Box) Amino acid sequences of Aβ peptides Aβ40 and Aβ42, with enzyme cleavage sites indicated. Polar residues are underlined. Part of the sequence of the amyloid precursor protein (APP) from which Aβ is cleaved by the secretase enzymes, is also shown above, with red residues (numbered) being sites of substitution in the mutants listed in Table 3.2.

Figure 3.5. The numbering scheme is required when mutants are discussed, as summarized in Table 3.2. The peptide N-terminus is created via cleavage by β-secretase in the extracellular domain of APP. The enzyme β-secretase is an integral membrane aspartyl protease encoded by the β-site APP-cleaving enzyme 1 gene (*BACE1*). The C-terminus of Aβ results from intramembrane cleavage by γ-secretase, which is a protein complex involving presenilins and other proteins. A third enzyme, α-secretase, cleaves between amino acids 16 and 17 in Aβ, thus hindering fibrillization. As can be seen from the sequence in Figure 3.5, the polar residues (underlined) are concentrated at the N-terminus in the two Aβ peptides.

The structure of Aβ has been investigated in detail and it is known that there is a β strand-turn-β strand configuration of the peptide forming a horseshoe-shaped strand, with the strands stacking upon one another to form the fibrils. Figure 3.6 shows the structure of Aβ42, determined from solution NMR experiments.

Of the two forms of Aβ, Aβ42 has a stronger aggregation propensity and forms fibrils more rapidly. This correlates with its greater observed neurotoxicity. Synthetic Aβ40 and Aβ42 are available from a large number of commercial suppliers, as are some animal-derived Aβ peptides.

A number of mutants of Aβ and the APP protein are implicated in specific variants of FAD listed in Table 3.2; also listed are the effect on APP and Aβ production. The residue numbering system of APP with respect to Aβ is shown in Figure 3.5. Many of these mutants can be expressed recombinantly in transgenic mice, used as models in many studies.

3.3.2   Treatments for Alzheimer's Disease

There are as yet no drugs that can cure AD. The only drugs currently available do not reverse or inhibit the progression of AD but may ameliorate the development of symptoms. The current standard of care

Table 3.2   Effect on Aβ of APP mutations.

Name/FAD variant	Mutation	Effect on APP	Effect on Aβ
APP-717 (London)	V717F/G/I	Differential γ-secretase cut	Aβ42:Aβ40 ratio increased
APP-670/671 (Swedish)	K670N and M671L	Increased β-secretase cut	Increased Aβ40 and Aβ42 in plasma
APP-692 (Flemish)	A692G	Decreased α-secretase cut?	Decreased Aβ40 and Aβ42 in media, decreased Aβ aggregation, Aβ42:Aβ40 ratio increased
APP-693 (Dutch)	E693Q	Unclear	Decreased Aβ42 in media, increased Aβ aggregation, Aβ42:Aβ40 ratio decreased
APP-693 (Arctic)	E693G	Unclear	Decreased Aβ40 and Aβ42 in plasma, Aβ42:Aβ40 ratio decreased
APP-693 (Italian)	E693K	Unclear	Decreased Aβ42 in media, Aβ42:Aβ40 ratio decreased
APP-694 (Iowa)	D694N	Unclear	Enhanced fibrillization of Aβ40
PS1-FAD mutations	M139I, H163A, and others	Differential γ-secretase cut	Aβ42:Aβ40 ratio increased
PS2-FAD mutations		Differential γ-secretase cut	Aβ42:Aβ40 ratio increased
Trisomy 21 (Down's syndrome)		Increased APP production	Aβ40 and Aβ42 increased
Apolipoprotein E4		Competes for LDL (low density lipoprotein) receptor-related protein (LRP)	Increased Aβ aggregation

for mild-to-moderate AD includes treatment with acetylcholinesterase inhibitors, to improve cognitive function, and with memantine, an NMDA antagonist. NMDA, as mentioned in the preceding section, is a neurotransmitter. Another important neurotransmitter target is acetylcholine, with therapies being based on inhibitors of the enzyme acetylcholinesterase, which catalyses the breakdown of acetylcholine. Acetylcholinesterase inhibitors include galantamine, an alkaloid available commercially with various trade names. Although developed as an acetylcholinesterase inhibitor, galantamine also acts to inhibit Aβ aggregation. It is also known

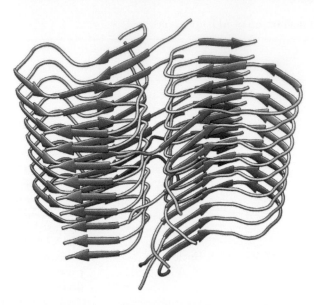

Figure 3.6 Structure of Aβ42 amyloid fibrils (pdb file 5KK3).

as an allosteric modulator of nicotinic receptors. Memantine is thought to function therapeutically as a channel blocker of NMDA receptors, and it also attenuates calcium influx into neurons.

Much research activity has been, and is being, devoted towards the development of therapies for AD. Since the 2000s, passive immunization based on antibodies has attracted considerable potential as an approach to treat AD. A series of monoclonal antibodies (raised against aggregating fragments of Aβ) were developed by pharmaceutical companies to the stage of phase II/III clinical trials. Unfortunately these were unsuccessful due to the lack of significant benefit and/or observed serious side effects. Active immunization using Aβ or its fragments directly has also not shown success.

Another promising avenue for therapeutics has been the development of secretase inhibitors. As shown in Figure 3.5, different secretase enzymes cleave Aβ from its precursor protein APP and also cleave within the sequence. It has proved difficult to identify small-molecule inhibitors of β-secretase (*BACE1* gene) with favourable pharmacokinetic characteristics. Statins may be used to inhibit cholesterol biosynthesis and the expression of *BACE1* and ultimately Aβ production. Cholesterol biosynthesis also affects binding of apolipoproteins (e.g. the *APOE* ε4 allele that is a signature of sporadic AD).

A greater number of compounds have been developed as potential inhibitors of γ-secretase. A complicating factor here is that other ligands are also γ-secretase substrates, leading to potential problematic side effects. However, certain nonsteroidal anti-inflammatory drugs (NSAIDs) such as ibuprofen can modulate γ-secretase cleavage without blocking Notch cleavage. The Notch transmembrane proteins play a key role in signalling between neurons. Secretase inhibitors have been developed to the stage of clinical trials (as yet without success) and this is still an active area of research activity. A less-explored route to the development of AD therapeutics has focused on the α-secretase cleavage of Aβ, although some studies are consistent with reduced Aβ formation by stimulation of α-secretase activity, for example using ADAM metalloproteases.

Many molecules have been investigated as potential Aβ aggregation inhibitors. These work by many mechanisms, for example by disrupting the β-sheet hydrogen bonds that characterize amyloid fibrils, or by influencing electrostatic interactions or the effects of species such as metal ions associated with amyloid aggregates (metal ion chelators). Aβ contains a metal binding motif involving histidine residues H6, H13, and H14 (the sequence is shown in Figure 3.5), and the Aβ peptides are known to form complexes with copper, iron, and zinc ions. Hydrogen bond disrupters include polyphenols that have attracted particular attention, including molecules such as tannic acid and epigallocatechin gallate (EGCG). These have been shown to reduce Aβ cytotoxicity in cell-based assays, and can disassemble mature Aβ42 fibrils, although these compounds are not able to cross the blood–brain barrier (BBB). Other biologically derived polyphenols including curcumin and rosmaric acid have also been examined as Aβ aggregation inhibitors, or as disaggregators of pre-formed fibrils. These and other flavonoids also have antioxidant properties that may be the basis of an alternative or additional mode of activity. The Amyloid-binding dyes (examples are shown in Figure 3.23) have been used as the basis of another class of Aβ aggregation inhibitor. Compounds that can block Aβ-induced channel formation in lipid membranes have also been examined as alternative aggregation inhibitors. It has been suggested that inhibitors of the interaction of Aβ with GAGs (glycosaminoglycans) could be promising aggregation inhibitors. GAGs or proteoglycans are associated with AD since sulfated GAGs such as heparan or chondroitin sulfate are present in neuritic plaques, neurofibrillary tangles, and vascular amyloid deposits. Binding of some sulfated GAGs can prevent the proteolytic degradation of fibrillar Aβ. Heparan or heparan sulfate can accelerate the fibrillization of Aβ *in vitro*. Peptidomimetic molecules (see Section 2.7) have also attracted attention as potential amyloid aggregation inhibitors, often being based around the KLVFF core aggregating sequence (Figure 3.5); for example D-amino acid variants and inverse sequences and others have been examined, along with sequences of this peptide with β- and γ-amino acids (which cannot fit into the hydrogen bonding pattern of the native residues and so can potentially disrupt fibrillization). Lipid-based small molecule inhibitors have also been developed, because the presence of lipid membranes may accelerate Aβ fibrillization, Aβ deposition being initiated in a plasma membrane-bound form. A particular focus has been on the interaction of Aβ with phosphatidylinositol since this causes a dramatic increase in fibril growth. This can be inhibited using headgroups from other members of the phosphatidylinositol family.

Proteins including chaperone proteins have been shown to bind to Aβ and to inhibit fibrillization/oligomerization. Polymeric and inorganic nanoparticles have also been developed that also show Aβ inhibition properties. However, proteins and nanoparticles face the significant challenge of the blood-brain barrier (BBB), making them less attractive as front-line candidate therapeutics for AD.

Many other studies report other small molecule inhibitors, nanoparticles, and composite systems that show activity against Aβ toxicity in cell studies and/or which show Aβ fibril or oligomer disruption. As yet, none of these have led to successful therapies (or even clinical trials with very few exceptions).

As well as treatment, the accurate diagnosis of AD is of great importance. This may be done using imaging methods such as MRI (magnetic resonance imaging), fMRI (functional MRI), or PET (positron emission tomography). Recently, sensitive ligands that bind Aβ peptides have been developed for PET imaging. As well as these imaging methods, which require access to large facilities in hospitals, methods to diagnose Aβ based on levels of the Aβ peptides in cerebrospinal fluid (CSF) have been established. However, CSF is obtained through the complex procedure of lumbar puncture (spinal tap) and so more straightforward tests based on blood plasma levels are being developed. Antibodies to Aβ may also be assayed using lab methods such as ELISA (enzyme-linked immunosorbent assay). Biomarkers associated with inflammation have also been investigated, in particular proinflammatory cytokines. Other disease-related biomarkers include ubiquitin and biomarkers related to cellular senescence such as the conformational state of protein p53 (involved in DNA repair or apoptosis) or telomere shortening. Finally, there are biomarkers associated with cerebrovascular damage.

3.4 MECHANISMS AND KINETICS OF AMYLOID AGGREGATION

The early stages of amyloid formation follow a nucleation-and-growth mechanism. Amyloid fibrils grow above a critical concentration (threshold), above which there is a thermodynamic driving force for aggregation.

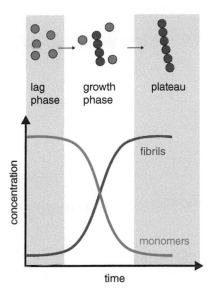

Figure 3.7 Concentration of species versus time, showing lag and growth phases during amyloid aggregation.

After the initial slow nucleation event, fibrils grow more rapidly. Amyloid aggregation is characterized by an initial lag phase (Figure 3.7), which indicates that the process is highly cooperative. This cooperativity is due to secondary nucleation events where existing amyloid fibrils act as catalysts for the production of new nuclei which then grow themselves. After the lag phase, the concentration of fibrils grows with a sigmoidal shape (Figure 3.7) as a function of time, along with a concomitant decrease in monomer concentration. The sigmoidal shape as a function of time t is represented by the empirical equation:

$$y = y_0 + \frac{A}{1 + \exp[-k(t - t_{0.5})]} \tag{3.1}$$

where k is the apparent rate constant, $t_{0.5}$ is the midpoint time, and A is the amplitude. The lag time can then be defined as $t_{lag} = t_{0.5} - 1/2\,k$. Equation 3.1 is empirical and alternative equations rigorously derived from kinetic theories are presented below (Eq. (3.2)).

Although monomer-dependent secondary nucleation occurs in the lag phase, the rate of secondary nucleation is maximal in the growth phase (where the gradient of fibril concentration versus time is a maximum). The primary nucleation process is often constant. The secondary nucleation process is thought to be important in the generation of toxic oligomeric species, for example oligomers are found to be associated with plaques. If the oligomers were generated instead by primary nucleation, their

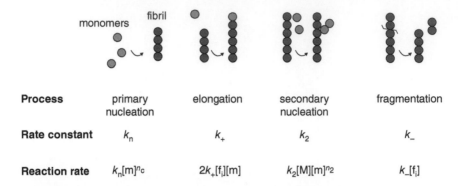

Process	primary nucleation	elongation	secondary nucleation	fragmentation
Rate constant	k_n	k_+	k_2	k_-
Reaction rate	$k_n[m]^{n_c}$	$2k_+[f_i][m]$	$k_2[M][m]^{n_2}$	$k_-[f_i]$

Figure 3.8 Microscopic processes that govern the rate of amyloid fibril formation, along with associated rate constants and reaction rates. Here [m] is the free monomer concentration, [M] is the total fibril concentration, and $[f_i]$ the fibril number concentration. n_C and n_2 are reaction orders for primary and secondary nucleation.

concentration would be low near the highly aggregated plaques. Natural chaperone molecules are those that inhibit secondary nucleation and reduce the toxicity associated with the aggregation process.

Figure 3.8 illustrates the different growth processes along with their associated rate constants and reaction rates. The overall expression for the rate equation is rather complex. To give an idea of magnitudes for Aβ42 in phosphate buffer, $k_n = 3 \times 10^{-4}$ M^{-1}, $k_+ = 3 \times 10^6$ M^{-1} s^{-1}, and $k_2 = 1 \times 10^4$ M^{-1} s^{-1} with $n_C = 2$ and $n_2 = 2$. These values can be used to determine the threshold concentration [M*] at which point the number of nuclei generated by secondary nucleation exceeds those produced by primary nucleation, [M*] $= 3 \times 10^{-8}$ M, which is a very low concentration.

In the absence of secondary nucleation and fragmentation, the kinetics of amyloid aggregation are described by the Oosawa model for one-dimensional assembly. The mass concentration of aggregates is described by the following integrated rate law which is derived from the rate equations:

$$M(t) = m(0) \left[1 - \text{sech}^{n_c/2}(\sqrt{n_c/2}\lambda t) \right] \tag{3.2}$$

Here $m(0)$ is the concentration of free monomers at time $t = 0$, sech denotes the hyperbolic secant function, n_c is the number of monomers per nucleus, and $\lambda = \sqrt{k_+ k_n m(0)^{n_c}}$ is a combined rate parameter. In the Oosawa model, the half time for aggregation is predicted to scale with the initial monomer concentration according to $t_{1/2} \propto m(0)^{-n_c/2}$. This can be compared with experimental measurements of aggregation kinetics for samples with different initial concentrations, as described in the discussion of Figure 3.9 following.

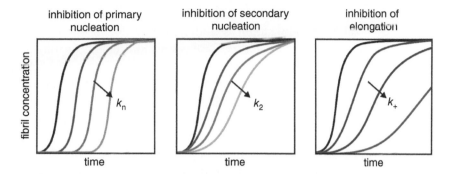

Figure 3.9 Schematic fibril growth curves, showing effect of the rate constants (decreasing in direction shown) for primary nucleation (k_n), secondary nucleation (k_2), and elongation (k_+).

The kinetic models to describe the case of the kinetics of fibrils undergoing fragmentation or secondary nucleation are more complex. In the case of fragmentation, the first-order equation for the aggregated mass is

$$M(t) = m(0) \left[1 - \exp\left(-2\frac{\lambda^2}{\kappa^2}\sinh^2\left(\frac{\kappa t}{2}\right) \right) \right] \quad (3.3)$$

Here $\kappa = \sqrt{k_+ k_2 m(0)^{n+1}}$, where n_2 corresponds to the nucleus size for secondary nucleation, in the simplest case.

In the case of monomer-dependent secondary nucleation, an accurate first-order solution from the rate equations is

$$M(t) = m(0) \left[1 - \left(\frac{B_+ + \lambda^2}{B_+ + \lambda^2 e^{\kappa t}} \frac{B_- + \lambda^2 e^{\kappa t}}{B_- + \lambda^2} \right)^{k_\infty^2/\kappa \tilde{k}_\infty} e^{-k_\infty t} \right] \quad (3.4)$$

Here $B_\pm = \kappa(k_\infty \pm \tilde{k}_\infty)$, $k_\infty = \sqrt{2\kappa^2/[n_2(n_2 + 1)] + 2\lambda^2/n_c}$ and $\tilde{k}_\infty = \sqrt{k_\infty^2 + \lambda^4/\kappa^2}$.

Fibril growth curves such as those shown in Figure 3.9 are commonly obtained from thioflavin T (ThT) fluorescence measurements of the kinetics of amyloid fibril growth. The shape of the growth curve depends on the initial monomer concentration and the rate of nucleation, which can be tuned by seeding. Figure 3.9 illustrates the effect of changing these conditions on the growth curves. Where secondary nucleation is dominant, the lag phase is only weakly dependent on the time taken for the initial nuclei to form, but instead depends on the rate of multiplication of nuclei and growth through elongation. The secondary nucleation process is thought to be important in the generation of oligomeric species. The lag phase is increased when primary nucleation is inhibited; on the other hand the growth rate curves are stretched out when elongation is inhibited by reducing the monomer concentration (or by added inhibitors, for example molecules targeting fibril inhibition, see Section 3.3.2).

The lag phase can be eliminated by addition of pre-formed aggregates, i.e. by seeding. As illustrated in Figure 3.10, seeding may be either homologous or heterologous, depending on whether the seed is from the same amyloid peptide or a different one.

Cross-seeding has been observed for several amyloid proteins This may be relevant *in vivo* due, for example, to cross-species transfer of prion proteins, which has been of concern in relation to the development of CJD from eating meat from cows with BSE (bovine spongiform encephalopathy). In addition, co-deposition of Aβ with PrP (prion protein) occurs in amyloid plaques observed in brain sections from AD patients. Cross-seeding with heterologous seeds often accelerates fibril formation. In the case of heterologous seeding, the efficiency of seeding seems to be correlated to the similarity in sequence between the host peptide/protein and the seed. Fibril morphology is also influenced.

For proteins, whether amyloid fibrillization results from partially folded intermediates containing β-sheet structures or from a fully denatured conformation has been studied. For most proteins, conditions that lead to partial unfolding favour fibrillization. As mentioned in Section 3.2, peptides are generally too short for folding intermediates to be relevant to amyloid formation. The extent of unfolding and ultimately the fibril morphology seem to depend on the level of denaturation. The temperature dependence of fibril aggregation rate constants for several peptides and proteins has been shown to follow Arrhenius-type behaviour.

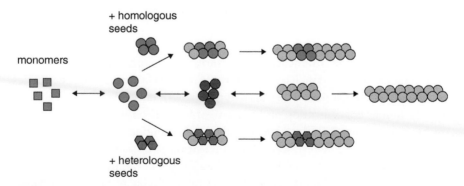

Figure 3.10 Homologous versus heterologous seeding. Colour scheme as for Figure 3.3.

Preparation conditions, such as shear forces during sonication and surface effects (nature of the surface of the vessel the sample is prepared in), affect both fibril morphology and toxicity. Other work shows that fibril morphology is strongly affected by preparation conditions, where aggregation is induced by pH, solvent, or heat, and can depend on the order in which preparation steps are performed. Also, especially for longer peptides such as Aβ peptides (with the inevitable synthesis challenges and low yield) there is unfortunately variability in the purity of the peptides obtained. There are often considerable differences among commercially supplied samples, depending on the source, for example whether synthesized (and synthesis procedures vary too when coupling fragments to prepare long peptides for instance, as discussed in Section 2.4.2) or obtained from recombinant protein expression. Altogether, this makes obtaining reproducible results on amyloid fibril kinetics quite challenging experimentally, although careful measurements have been performed and have yielded considerable insight, enabling the accurate testing of the kinetic models discussed above, for example.

3.5 TOXICITY AND RELEVANCE TO DISEASE

3.5.1 Oligomers

Oligomers are believed to be toxic species formed by many amyloid-forming peptides and proteins, rather than fibrils (since much work has been done with Aβ, this is used as an exemplar in the following). Evidence for this comes from several experiments on disease-related and non-disease-related peptides.

(i) *In vivo* and cell culture experiments show that oligomers are neurotoxic.
(ii) Oligomeric forms of Aβ (such as dodecameric aggregates) disrupt learning behaviour using mouse and rat models.
(iii) Anti-Aβ antibodies isolated from immunoglobulin strongly disrupt fibrillization.
(iv) Polyclonal antibody experiments indicate that antibodies suppress the toxicity of soluble oligomers whereas there is no antibody response to mature fibrils.

On the contrary, evidence that fibrils are not the toxic species includes the following observations:

(i) No clear correlation is found between the extent of fibril formation or amyloid load and AD.
(ii) To date, therapies to remove amyloid aggregates (plaques) have not improved AD symptoms.
(iii) Amyloid fibrils are found in the brains of patients who have not died from AD.

As well as their potential role in disease, soluble oligomers are of interest as biomarkers for early diagnosis and for assessment of therapeutic efficacy in trials. Due to their demonstrated neurotoxicity, oligomers are potential therapeutic targets. In particular, Aβ oligomer-specific antibodies were intensively researched for AD treatments, but as discussed further in Section 3.3.2, this has not as yet led to a clinically successful drug. Similar approaches have been extended to other amyloid proteins, including α-synuclein and tau.

Specific oligomer forms have been observed with defined numbers of aggregated monomers in globular clusters (ring-like or annular structures have also been reported). Obtaining reproducible and stable oligomer samples is important for many tests on amyloid systems, for example testing potential therapeutic molecules. Methods to prepare particular oligomer formulations are available, including specific solution processing of Aβ, from a starting HFIP (hexafluroisopropanol) solution (which generates monomers) and chromatographic separations, or chemical cross-linking when defined oligomeric aggregation states are achieved. Protofilament structures may also be defined as oligomers. Oligomers may be detected using specific antibodies and SDS-PAGE (sodium dodecyl sulfate–polyacrylamide gel electrophoresis) or Western blotting. Alternatively, antibody-free analytical methods including size exclusion chromatography or sedimentation or ion-mobility mass spectrometry may be employed. Chemical cross-linking methods have also been developed to trap specific oligomeric species for characterization or application, especially for smaller oligomers which are less stable transient species.

The observation that the number of oligomers is reduced when fibrils form suggests that most oligomeric species are on the pathway to fibrillization (Figure 3.3). This has been observed for nearly all amyloid-forming peptides and proteins. This suggests that oligomers are obligate structures en route to fibril formation. Oligomers show a considerable variation in their structures and β-sheet content, which is generally much lower than that of fibrils (they show weak binding of amyloid-diagnostic dyes including ThT and Congo red). The variability in structure is demonstrated by the fact that a vast number of antibodies are available that are sensitive to different oligomeric species. However, some antibodies seem to recognize common features of certain types of oligomer, across different proteins and peptides.

The annular pore-like oligomers observed for several amyloid proteins and peptides suggest a mechanism of action involving membrane pores or channels. Annular pore-like oligomers have been observed during the aggregation of a number of disease-associated and non-disease-related amyloid systems. Furthermore, addition of oligomers has been shown to lead to pore formation in lipid membranes. As mentioned in Section 3.3.1, the mechanism of Aβ cytotoxicity may be due the formation of membrane pores or channels arising from exposure of hydrophobic regions in the peptides. Positive charge on peptides that enables interaction with negatively charged lipid membranes may also be important. It has been shown that Aβ42 oligomers bind to cell membranes and cause cytotoxicity under conditions in which mature fibrils do not form. Aβ forms pores in lipid membranes that contain multimers of the protein, as revealed by AFM. Evidence for pore formation in $vivo$ has also been obtained via TEM on neuronal cell membranes. Uptake of Ca^{2+} across the ion channels leads to neuronal degeneration in a dose-dependent and time-dependent manner, and ultimately cell death. Changes in the calcium level and the morphology of cultured cells are sensitive to the aggregation state of Aβ42. Aβ-pore formation leading to an increase in intracellular calcium has been linked to depletion of synaptic vesicles and, hence, blocked neurotransmission.

A cautionary note needs to be sounded regarding these observations. The studies on oligomers are not performed in $situ$, instead the experiments rely on in $vitro$ cell studies or in $vivo$ measurements in which the oligomer aggregation state may change after injection into the brain but before toxicity measurements are made, which may be performed hours or days after application. Oligomerization of Aβ occurs intracellularly and oligomers have been detected in neural and non-neural cell lines. However, there are few studies that have examined the structural and biochemical diversity of oligomers in $vivo$ and the mechanisms of oligomer production and their subsequent fate in the body are still relatively unexplored.

Post-translation modifications (PTMs) such as phosphorylation, lipidation, and nitration may regulate the equilibrium between different oligomer/fibril states and so have been suggested as potential targets for therapies. Clearly, such structural modifications can be expected to significantly influence peptide aggregation. To date, most toxicity studies do not consider PTMs.

3.5.2 Amyloid Formation Is Not Sequence Specific

Evidence that formation of amyloid fibrils is a common state for many if not all proteins comes from several types of experiments. First, fibrils can be induced to form by partial denaturing of proteins not involved with any disease or using *de novo* designed peptide fragments. Secondly, amyloids can be induced to form by seeding with fibrils of the same, related, or unrelated proteins.

The cross-β structure (Section 3.2) is a common feature for amyloids formed by many different proteins and peptides. Cross-β structures are observed independent of sequence or side-chain type for polyamino acids, including poly(L-lysine), poly(L-glutamic acid), or poly(L-threonine). This is quite distinct from the protein folding process, which depends on the specificity of side-chain interactions. However, in certain conditions (of temperature or pH) aggregation in amyloid fibrils can be overcome by specific side-chain interactions, which may lead to kinetically favourable states or may destabilize fibrillar aggregates. It thus appears that fibril formation is due to the common main-chain polypeptide backbone whereas folding is due to specific interactions of the side chains.

Amyloids also have the common property that they are stained by Congo red and thioflavin dyes. This may simply reflect the common cross-β structure. One antibody (called 6E10) derived from soluble oligomeric intermediates of Aβ also recognizes oligomers from a range of other proteins and peptides. Recognition is not observed for low M_w or fibrillar Aβ species. This indicates that the antibody recognizes a common epitope in soluble oligomers.

3.6   FIBRILLIZATION OF SMALL PEPTIDES

Many short peptides that contain both hydrophobic and charged residue sequences will assemble into β-sheet structures under certain conditions. Indeed, in contrast to long peptides where the α-helix structure is more common, the β-sheet structure is the most common secondary structure for conformationally ordered short peptides. Of course, very short peptides do not have enough residues to form α-helical repeats but peptides as short as tripeptides can form β-sheet structures, and even dipeptides form β-sheet-like structures.

As well as a large number of *de novo* designed short peptides which have been shown to form β-sheet structures, due to the intense research interest in amyloid formation of relevance to protein misfolding diseases, much work has been done using model amyloid-forming peptides. One widely studied amyloid-forming peptide is GNNQQNY from the yeast prion protein Sup35. There are many studies on its amyloid properties and a crystal structure determination has revealed the formation of a steric zipper in which the side chains are intercalated within the cross-β structure (Figure 3.11).

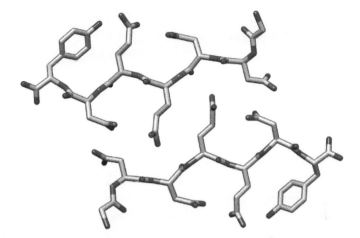

Figure 3.11 Steric zipper structure obtained from single-crystal X-ray diffraction structure of peptide GNNQQNY (pdb file 1YJP), showing the zipping or interdigitation of side chains.

The VQIVYK sequence from tau (residues 306–311) forms a β-sheet structure and the crystal structure also suggests that the β-sheets are stabilized by steric zippers. The peptide NFGAIL from hIAPP (human islet amyloid polypeptide also known as amylin, Table 3.1) has been extensively used as a model amyloid-forming peptide. Other examples of steric zippers, which exclude water from the interior of amyloid fibrils, include phenylalanine zippers. The peptides KLVFF and KLVFFAE from the Aβ peptide (residues 16–20 and 16–22, respectively) are believed to be core aggregating sequences. Indeed, peptides incorporating these fragments can bind the full-length peptide and prevent fibrillization. Fragments based on KLVFF have been designed to inhibit Aβ40 and Aβ42 fibrillization, as have related peptidomimetics (Section 3.3.2). Rational design principles based on the knowledge of the pentapeptide binding sequence led to the preparation of LPFFD, a β-sheet breaker peptide. This peptide incorporates proline, known to hinder β-sheet aggregation (see Figure 1.9), and was found to reduce amyloid deposition *in vivo* (rat model) and to disassemble preformed fibrils *in vitro*. However, as mentioned above, Aβ fibrillization inhibition has not as yet proven to be an approach that has yielded a viable therapeutic. Within KLVFFAE (and the β-breaker peptides) is the FF dipeptide which has been proposed as a driver of aggregation. This dipeptide itself is able to assemble into fibrils (and nanotubes) due to the formation of β-sheet-like structures driven by strong π-stacking interactions (Section 3.10).

3.7   BIOLOGICAL FUNCTIONAL AMYLOID AND BIOENGINEERING APPLICATIONS OF AMYLOID MATERIALS

Nature has evolved amyloids for a number of functional and structural uses. Fungi produce hydrophobins, which are small proteins/large peptides that form hydrophobic coatings, enabling fungi to spread on hydrophobic surfaces. Hydrophobins also enable the formation of filamentous structures termed hyphae by reduction of surface tension at the air–water interface (Figure 3.12); these develop into aerial structures including spores and fruiting bodies. Chaplins produced by Streptomyces bacteria also reduce surface tension, facilitating the development of hyphae that grow as part of the spore formation process. Amyloid in fungi has other roles, including regulation of nitrogen catabolism (URE2p amyloid) and yeast cell fusion (HET-s protein). Enterobacteria including *Escherichia coli* and *Salmonella* species produce curli which are amyloid fibres (of the CsgA protein) produced as part of the extracellcular matrix (ECM) in biofilms as the bacteria form spreading colonies.

Many other proteins form functional amyloids, with biological roles in the formation of biofilms (see Section 4.4), adhesive pili and filaments of bacteria, and the protective shells of insect and fish eggshells. Since these are based on protein amyloids, detailed discussion is outside the scope of this book. The melanosome in humans produces the pigment melanin. It contains amyloid formed from protein Pmel17. The amyloid has a structural role, being involved in sequestration of toxic intermediates during melanin synthesis. Melanin is produced from the oxidation of tyrosine by the enzyme tyrosinase, which first produces L-dopaquinone, followed by polymerization.

Silk-like peptides are attracting interest due to their potential excellent mechanical and structural properties. Silk forms amyloid-like β-sheet

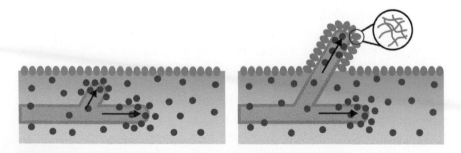

Figure 3.12   Formation of filamentous hyphae by fungi and Streptomyces bacteria. Amyloid aggregates (green) of hydrophobins or chaplins reduce surface tension at the air–water interface and these proteins in monomeric form (red) facilitate growth of filaments.

structures. Silk has a semicrystalline structure with crystalline domains embedded in an amorphous matrix. Silk from the *Bombyx mori* silk moth can be considered to be a multiblock copolymer. The crystalline domains contain repetitive $(GAGA)_n$ domains forming stacked β-sheets as well as GS and GT dipeptide sequences. On the other hand, dragline spider silk contains poly(alanine) repeats.

Unexpectedly, so-called metabolite amyloids are observed for dipeptides such as FF and FW, which form fibril and nanotube (Section 3.10) structures in aqueous solution. Even more remarkably, phenylalanine itself can form fibrils and this has been shown to have relevance to phenylketonuria (PKU), which is a disease associated with a mutation in the gene for the enzyme phenylalanine hydroxylase. Mice raise antibodies to phenylalanine fibrillar assemblies. Other amino acids, including tyrosine, have also been shown to form amyloid-like assemblies.

Peptide hormones (see Section 5.3) are stored in secretory granules in the form of amyloid structures and it has been noted that this may be correlated to the observation that peptide hormones are over-represented among amyloid-forming peptides, for example amylin, calcitonin, and atrial natriuretic factor (ANF) (Table 3.1) are all amyloid-forming hormones. Storage in the amyloid form ensures the efficient packing of the peptides at high density. In addition, provasopressin and vasopressin (see Section 5.3.4), form amyloids.

3.8 FIBRILS FROM α-HELICES

In general, peptide fibrils are formed from β-sheet assemblies. Collagen is an important counter-example; it forms an extended fibril-like structure with a right-handed triple helix structure. Collagen contains repeating GXY triplets where X and Y are often proline and hydroxyproline, respectively. Collagen is a protein; shorter peptides based on its sequence have been produced which show collagen-like structures.

It is also possible to design α-helical peptides to form fibrils through coiled coil assembly, based on the heptad of hydrophobic and polar residues HPPH-PPP (see Section 1.4). Complementary flanking ion pairs (E-K interactions), along with additional asparagine–asparagine pairing in the core, are used to create staggered heterodimer 'sticky ends' to promote the formation of long fibres; these are so-called self-assembling fibres (SAFs) (Figure 3.13). Fibres can also be obtained by conformational switching into β-sheet structures or α-helix-rich proteins or peptides can be 'denatured' to form amyloid.

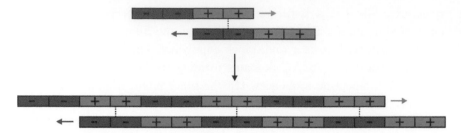

Figure 3.13 Schematic of self-assembling fibre peptides due to complementary electrostatic interactions within heterodimers, also stabilized by interactions between asparagine residues indicated by the dashed lines. Each box indicates a heptad repeat.

3.9   PEPTIDE HYDROGELS AND TISSUE SCAFFOLDS

A large number of peptides will form hydrogels under appropriate conditions (sufficiently high concentration, adjustment of pH, etc.), especially under conditions where fibril formation is favoured. Hydrogels are self-supporting structures (i.e. they have a yield stress) with variable stiffness that can be measured using rheological methods (Section 3.12.5). These hydrogels are useful in the development of peptide-based biomaterials such as scaffolds for cell growth in tissue engineering applications, or as slow-release systems. Some peptide therapeutics are formulated in gels for sustained release, especially for subcutaneous delivery. One example is discussed in Section 5.3.1, for example. Peptide-based molecules are also capable to act as organogelators, with applications in oil/pollutant recovery and catalysis, among others.

Peptide hydrogels are generally formed when peptide fibrils form a physically cross-linked sample-spanning network. The fibrils form a dense mesh structure, as shown schematically in Figure 3.14. Gels may form at sufficiently high concentration of peptide, and this may be facilitated by screening of charges on the peptide molecules by preparation in buffer or salt solutions. Alternative methods to promote hydrogel formation include pH variation, in particular pH increase (to around pH 8 or 9) followed by pH decrease (acidification). The initial pH increase favours peptide monomer formation due to electrostatic repulsion, whereas the pH decrease reduces the net peptide charge, favouring association. Gelation is often observed near the pI of a peptide. An alternative method to decrease the pH has been found to produce more homogeneous gels, which are more transparent. This is by slow pH decrease using the slow hydrolysis of the molecule glucono-delta-lactone to gluconic acid; this reaction is used in food processing when gradual hydrolysis is required. Gelation can also be promoted by heating, which can enhance

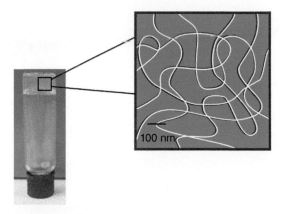

Figure 3.14 A peptide hydrogel in a vial with overlaid schematic of microscopic structure comprising a network of fibrils.

peptide solubility and/or denature peptides or proteins (favouring amyloid fibril formation).

Enzymatic processes have also been used to produce peptide gels, in which the enzyme produces a less soluble version of the peptide molecule which is a gelator (for example by dephosphorylation, or reverse hydrolysis). Figure 3.15 shows a schematic of these processes for dipeptides N-terminally modified with Fmoc (fluorenylmethyloxycarbonyl) or naphthalene that have been used as model peptide-based gelators. These reactions have been demonstrated with these and related bulky N-terminal groups in many cases and for some classes, e.g. Fmoc dipeptides (whether formed enzymatically or prepared directly), gelation is observed with many combinations of residues (especially those containing one or more aromatic residues). In the case of the reaction shown in Figure 3.15c, for example, gelating Fmoc dipeptides can also result from thermolysin-catalysed condensation of an Fmoc amino acid and an aromatic amino acid.

Many types of peptide molecules will form fibrous gels if the concentration is high enough, although it is not possible to predict *a priori* whether a peptide will form a gel or not; attempts are being made to account for gelation based on solubility parameter-based models. As mentioned above, addition of bulky substituents such as Fmoc or naphthalene groups is a current popular method to generate gels.

As well as forming hydrogels themselves under appropriate conditions, certain peptides are incorporated within polymeric hydrogels (water-based polymer gels) that are of interest as scaffolds for cell culture. In particular, cell adhesion peptides may be incorporated (for example as side chains on the polymers) in order to stimulate cell attachment and growth, or

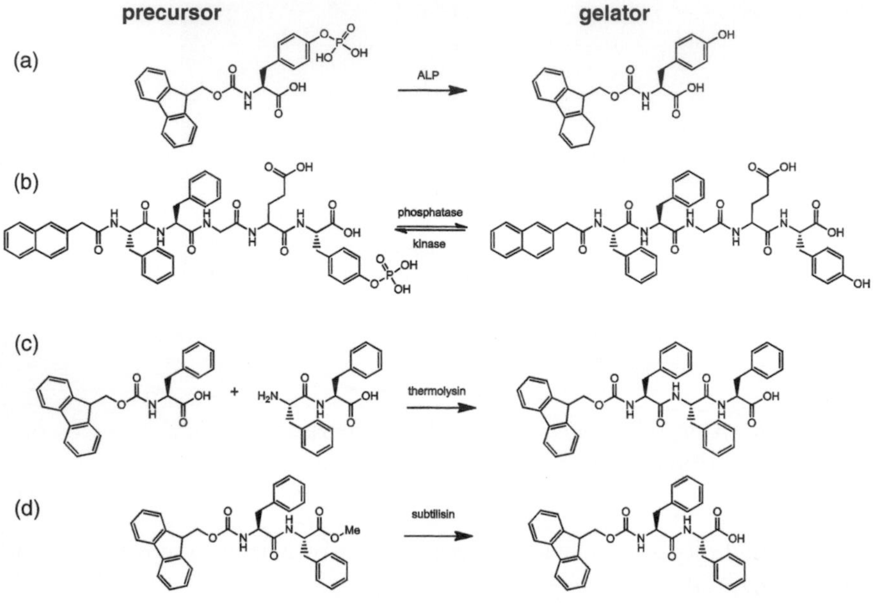

Figure 3.15 Examples of enzymatic reactions using Fmoc- and naphthalene-modified small peptides. (a) Dephosphorylation using alkaline phosphatase (ALP) reduces the solubility of Fmoc-phosphotyrosine leading to a gelating Fmoc-tyrosine peptide. (b) A short naphthalene-tetrapeptide also uses a phosphotyrosine to reversibly form a hydrogel. (c) Reverse hydrolysis of an Fmoc-amino acid and an aromatic dipeptide using thermolysin leads to condensation of an Fmoc-tripeptide hydrogelator. (d) Ester hydrolysis using subtilisin leads to the creation of hydrogelator Fmoc dipeptides.

differentiation into distinct cell types in tissue engineering applications. Such peptides are also valuable in the development of new materials that facilitate wound healing. Hydrogels constitute a moist environment with sufficient oxygen supply that helps cells to regenerate. They prevent bacterial growth due to the narrow gel mesh size (typically 0.1–5 μm), which acts as a mechanical barrier for bacteria. Hydrogels can also absorb and retain the wound extrudate, enabling the proliferation of fibroblasts (connective tissue cells which produce ECM) and the migration of keratinocytes which form the barrier in the healed epidermis. Some key cell adhesion motifs are listed in Table 3.3.

Cell adhesion to a surface such as ECM requires multiple cues, as shown in Figure 3.16. The RGD tripeptide is the minimal unit of a cell-adhesive domain present in adhesion proteins such as fibronectin, fibrinogen, and vitronectin, which are components of the ECM that all contain integrin ligands. RGD is also presented as an adhesion recognition sequence in other ECM

Table 3.3 Cell adhesion and related motifs.

Sequence/Name	Source	Activity
RGD(S)	Fibronectin	Integrin cell adhesion motif
PHSRN	Fibronectin	Synergy domain
REDV	Fibronectin	Integrin cell adhesion motif
YIGSR	Laminin	Integrin cell adhesion motif
IKVAV	Laminin	Integrin cell adhesion motif
DGEA	Collagen (type I)	Integrin cell adhesion motif
KRSR	Designed based on consensus repeat	Heparan binding and osteoblast adhesion
GFOGER/GFPGER	Collagen (type I and type IV)	Integrin cell adhesion motif

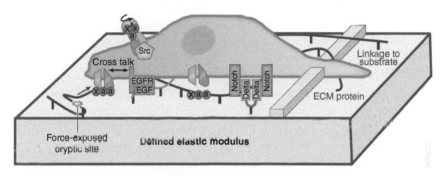

Figure 3.16 There are multiple cues for cell adhesion with the ECM. These include integrins (labelled α, β), cell adhesion motifs Xaa (e.g. RGD), growth factors and their receptors (e.g. epidermal growth factor receptor, EGFR), and Notch transmembrane cell signalling receptors. Delta is a morphogenic biochemical factor involved in ligand clustering. Src is a mechanotransductive tyrosine kinase associated with focal adhesions.

proteins, including laminin and some types of collagen. It is widely used as a cell adhesion motif in a vast range of biomaterials research, interacting with the transmembrane integrin $\alpha_v\beta_3$ and $\alpha_v\beta_5$ receptors, and was originally identified as a key factor in angiogenesis. The related tetrapeptide RGDS also exhibits high cell adhesion activity via binding to the integrin cell-surface receptors and, like the shorter RGD variant, is widely used to encourage cell growth in synthetic biomaterials. The tetrapeptide sequence and more especially the tripeptide RGD subunit are widely employed in the development of bionanomaterials for applications in cell growth/differentiation or tissue scaffolding. The RGDS tetrapeptide has antithrombolytic activity due to the inhibition of platelet aggregation resulting from the fibrinogen recognition sequence. The origin of the efficacy of the RGD cell adhesion motif has been considered. It has a particular charge pattern of anionic D and cationic R separated by a neutral glycine residue. However, the properties of arginine (such as its ability to form bidentate hydrogen bonds with carboxylic acids) must be of particular importance since the analogous KGD sequence is known as a 'disintegrin' sequence, for example it is found in some snake venom peptide sequences. Disintegrins are short proteins/large peptides with activity as inhibitors of both platelet adhesion and integrin binding.

High affinity recognition by the $\alpha_5\beta_1$ integrin requires a so-called synergy peptide PHSRN, a sequence present in fibronectin. The REDV (and lesser known LDV) sequences are also cell adhesion motifs from fibronectin. They bind to the $\alpha_4\beta_1$ integrin. The REDV tetrapeptide bears obvious sequence similarity to RGDS. The KQAGDV sequence also promotes smooth muscle cell adhesion.

The YIGSR peptide is a cell adhesion domain from the ECM protein laminin. Both poly(RGD) and poly(YIGSR) have been reported to show anticancer activity. Multimeric YIGSR also inhibits angiogenesis and tumour growth of human fibrosarcoma cells in mice. The IKVAV peptide, also obtained from laminin, promotes cell attachment, spreading, and neurite outgrowth. A minimal sequence DGEA from type I collagen is able to recognize the $\alpha_2\beta_1$ integrin receptor. Peptides containing this sequence are able to inhibit adhesion of platelets to collagen mediated by $\alpha_2\beta_1$. For breast adenocarcinoma cells which use $\alpha_2\beta_1$ as a collagen/laminin receptor, adhesion of collagen and laminin was inhibited by DGEA-containing peptides. The KRSR motif was discovered as a consensus repeat from protein database analysis of bone-related adhesion proteins. The GFOGER domain (O is hydroxyproline, see Table 1.3), originally identified based on the sequence GFPGER within collagen I, is a high-affinity binding site to the integrins in collagen. Since the $\alpha_2\beta_1$ integrin receptor is involved in osteogenesis, the GFOGER domain has been incorporated into biomaterials designed to assist bone repair.

3.10   PEPTIDE NANOTUBES

3.10.1   Nanotube Structures

The most common form of peptide nanotubes are those formed by closure of helical ribbons (Figure 3.17a) into nanotubes. The nanotube wall comprises β-sheet layers of peptide molecules lying perpendicular to the tube axis. If the nanotubes self-assemble in water, then typically the surfaces of the nanotube will comprise charged residues while the interior of

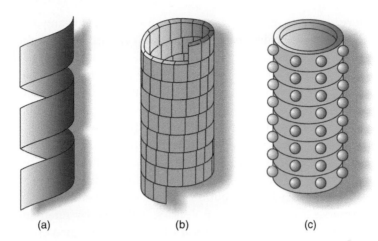

(a)   (b)   (c)

Figure 3.17 Three classes of observed peptide nanotube structures: (a) beta-sheet tapes wrapped into helical ribbon, (b) helical tubule from peptide subunits (bricks), (c) stacked cyclic peptides.

the nanotube wall will be composed of hydrophobic residues. The molecules may be perpendicular to the tube axis, or may be tilted. Figure 3.18 shows the closure of a nanoribbon into a nanotube in the case of a wall comprising a bilayer of molecules. Helical ribbon formation is driven by β-sheet curvature caused by packing and electrostatic interactions. The helical ribbon state may coexist with twisted tape structures (see, for example, Figure 3.20).

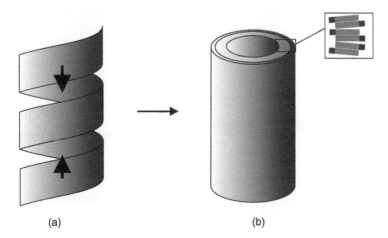

(a)   (b)

Figure 3.18 Closure of helical ribbons (a) into a peptide nanotube comprising a bilayer of amphiphilic peptide molecules with (b) a hydrophobic interior (light blue) and hydrophilic exterior (dark blue).

The structure shown in Figure 3.17a has been observed for amyloid peptide fragments, surfactant-like peptides, and lipopeptides. Remarkably, the dipeptide diphenylalanine is able to self-assemble into nanotubes in aqueous solution, based on structures resembling β-sheets that form the nanotube wall. These nanotube structures have remarkable stiffness and optoelectronic properties, have been shown to behave as semiconductors and quantum dots, and show effects such as piezoelectric activity, quantum confinement, photoluminescence, and photosensitization.

The type of nanotube represented in Figure 3.17b is only observed for a small number of peptides, one example being the cyclic octapeptide lanreotide (see Section 5.3.1). The building blocks of the nanotube wall in this case comprise a bilayer of the molecules, stabilized by interdigitation of hydrophobic side chains in the interior of the wall. Designed cyclic peptides that comprise alternating D- and L-residues that can form stacked ring nanotube structures are shown schematically in Figure 3.17c and in more detail in Figure 3.19. These structures are stabilized by networks of hydrogen bonds running along the nanotube wall.

3.10.2   Mechanism of Nanotube Formation

The mechanism of nanotube formation can be imaged in real time using AFM. Figure 3.20 shows an example of AFM height images taken during the aggregation process of an Aβ peptide-based fragment-based peptide which forms twisted tapes, then ribbons, and then finally, after low temperature incubation, nanotubes. As with other peptide aggregation processes,

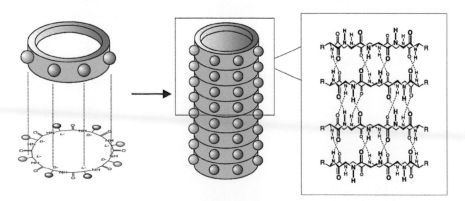

Figure 3.19   Stacking of alternating D,L-cyclic peptides to form nanotubes, stabilized by a hydrogen bonding network shown on the right. Residues are shown as coloured beads.

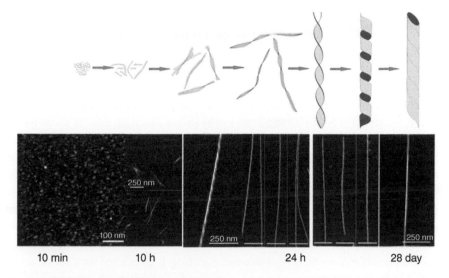

Figure 3.20 Bottom: AFM height images during the time-dependent aggregation of peptide βAβAKLVFF (βA: β-alanine, see Table 1.3) during incubation at 25 °C. After 10 min small oligomers are observed, after 10 h nucleation of protofilaments is evident, but after 24 h twisted tapes and helical ribbons are formed. Finally, after 28 days of incubation at 4 °C, closed nanotubes have developed. Top: schematic of the corresponding structures.

the assembly can also be deduced from spectroscopic methods such as fluorimetric determination of aggregation. Also, since as discussed above, peptide nanotubes are often based on β-sheet structures, assembly leads to characteristic bands in CD (circular dichroism) and FTIR (Fourier transform infrared) spectra, and in some cases also bind the 'amyloid' sensitive dye ThT (Section 3.12.1)

3.11 PEPTIDE AND PEPTIDE CONJUGATE ASSEMBLIES

3.11.1 Lipidated Peptide Assemblies

As mentioned in Sections 2.8 and 2.9, nature has evolved many lipidated peptide structures which are created post-translationally and, of course, lipopeptides can be readily synthesized by terminal (especially N-terminal) lipidation. Lipidation confers extended stability *in vivo* due to reduced proteolysis and stabilization by facilitating binding to serum albumin in circulation.

Lipopeptides are also known as peptide amphiphiles (PAs). They have been observed to self-assemble into several nanostructures including fibrils, micelles, nanosheets, vesicles (Figure 3.21), and nanotubes (discussed in Section 3.10). The most commonly observed structures are fibrils and nanosheets, which are lamellar bilayer structures. The terms 'fibrils', 'fibres', 'nanofibres', and 'nanofibrils' are used interchangeably. Nanotapes are extended structures containing layers of β-sheets but with a finite sheet width and extended length. As evident for example in Figure 3.20, nanotape structures are usually twisted. As discussed earlier in this chapter, amyloid fibrils are also typically twisted. The degree of twisting of nanotapes and peptide fibrils has been described in different models based on (i) the torsional elastic energy of stacks of β-sheets balanced against the attractive interaction energy (dispersion energy) between β-sheets, or (ii) the torsional elastic energy balanced by the electrostatic energy associated with the charged sheets or sub-filaments. The degree of twisting is predicted as a function of tape thickness and width, or as a function of ionic strength, respectively, in these two types of model.

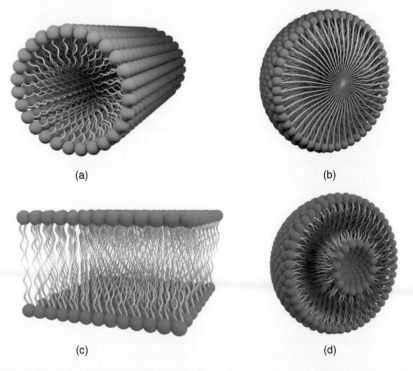

(a)

(b)

(c)

(d)

Figure 3.21 Lipopeptide self-assembled nanostructures. (a) Fibrils, (b) micelles, (c) nanosheets (bilayer lamellae), (d) vesicles. The hydrophilic peptide is represented as a blue sphere 'head group', the hydrophobic lipid 'tail group' as a green chain.

The self-assembled structure formed by a lipopeptide is dependent on a combination of factors, including molecular shape, hydrogen bonding, and electrostatic effects. The effect of shape can be rationalized by adopting the surfactant packing parameter model, which describes the packing of molecules represented as simple geometrical objects such as cones or cylinders. This model is widely used to qualitatively describe the morphologies observed for surfactants. In the case of lipopeptides, strong intermolecular interactions, in particular hydrogen bonding, mean that a simple geometrical picture is not sufficient to account for the shape of the self-assembled nanostructure and it is sometimes difficult to predict which structure will form for a given lipopeptide molecule. However, computer simulation studies predict phase diagrams similar to that shown in Figure 3.22, which shows the favourable β-sheet tape structures when hydrophobic effects are relatively weak but hydrogen bonding is strong (molecules do not aggregate if both energies are weak). Strong hydrophobic interactions favour the formation of micelles when the hydrogen bonding energy is low, or favour extended cylindrical fibrils when the hydrogen bonding energy is also high. The effect of electrostatic and other interactions further complicates the picture. Lipopeptides bearing short charged sequences may form micelles, those containing longer peptide sequences are likely to form hydrogen bonded β-sheets that may self-assemble into fibrils. Aromatic stacking interactions typically favour extended structures such as fibrils or nanotubes.

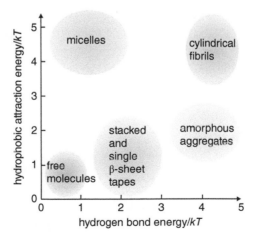

Figure 3.22 Schematic phase diagram for lipopeptide assemblies in terms of the energies of hydrophobic and hydrogen bonding interactions. Energies are in molecular units (k is Boltzmann's constant and T is temperature). Source: Based on Velichko et al. 2008.

Lipopeptide assemblies such as vesicles or micelles may be useful as delivery vehicles to encapsulate, respectively, hydrophilic or hydrophobic cargo such as drug molecules. Lipopeptides that self-assemble into fibrils can be used to present bioactive peptide motifs at high density on the fibril surface. In addition, self-assembly may influence bioactivity where these self-assembled nanostructures are observed in biological media. Note that structures observed in water are not necessarily stable in media due to charge screening interactions by the salts in the media, etc. The assemblies may interact directly with cells or cell receptors, or bioactivity may be influenced by changes in concentration due to the molecule–aggregate equilibrium. This is an area of active research. There is evidence that self-assembled structures can be taken up by cells. Fibrillar structures generally have a longer circulation time *in vivo* as they are broken down more slowly than globular structures.

Lipidation (or other conjugations) can influence the secondary structure of a peptide, since its conformation may change due to steric or other constraints caused by packing into different aggregates. This can readily be studied by spectroscopic methods such as CD (see Section 1.5.2). Lipopeptide structures (and peptide and other peptide conjugate structures) can be imaged using electron microscopy (especially cryo-TEM) and AFM (Section 3.12.3). They can also be probed by small-angle scattering (SAS) (Section 3.12.2), which gives accurate information on the shape and dimensions of the nanostructures.

Lipopeptides are being investigated for a variety of biomedical applications, especially in regenerative medicine. They have been shown, for example, to stimulate the regeneration of neurons, cartilage, bones, nerves, and other tissue by presentation of peptide sequences on lipopeptide fibril surfaces that are based on motifs that bind to the corresponding biomaterial. This research should lead to products useful for biomedical practice within the next few years.

3.11.2   Polymer–Peptide Conjugate Assemblies

Polymer–peptide conjugates have been observed to form fibrils, nanotubes, micelles, and vesicles. Conjugates of hydrophobic peptides with hydrophilic polymers such as polyethylene glycol (PEG) typically form fibrils with a PEG corona. Vesicles have also occasionally been observed for polymer–peptide conjugates and block polypeptides. Nanotubes can be produced by grafting polymers as side chains onto cyclic peptides, with the cyclic peptides then stacking into nanotubes, as in Figure 3.17. PEGylated peptide assemblies are of considerable interest since PEG confers a so-called 'steric stabilization' effect (i.e. the PEG forms a barrier or coating around the peptide preventing it from being degraded), allowing longer *in vivo* circulation. PEGylated peptide hormone drugs (such as insulin) have been developed for this reason.

3.11.3 Self-assembling Peptides

Many peptides can assemble into 'amyloid' fibrils, and related extended nanotape structures are also commonly observed as β-sheet assemblies. Other structures, including vesicles and micelles, have been observed for classes of self-assembling peptide which include (i) surfactant-like peptides, (ii) peptide bolaamphiphiles, and (iii) ionic-complementary self-assembling peptides. Table 3.4 shows examples of sequences from these three classes. Surfactant-like peptides contain sequences of hydrophobic residues capped by hydrophilic residues, giving a surfactant-like structure with a hydrophobic tail and hydrophilic head. Peptide bolaamphiphiles comprise sequences with the same hydrophilic residue or short sequence on each side of a hydrophobic central sequence. Bolaamphiphiles have also been prepared with hydrophilic peptides capped on each side by lipid chains. Ionic-complementary peptides comprise distributions of charged residues (separated by neutral residues such as alanine). The charged groups have patterns $+-$ (type I), $++--$ (type II), $+++---$ (type III), or $++++----$ (type IV). The EAK16 and RADA16 in Table 3.4 are examples of type II and

Table 3.4 Examples of surfactant-like peptides, peptide bolaamphiphiles, and ionic-complementary self-assembling peptides.

Name/Sequence	Self-assembled structure
Surfactant-like peptides	
A_6R	Nanosheets/nanotubes
A_6K	Nanotubes
V_6D	Nanotubes/vesicles
R_4F_4	Nanosheets
Peptide bolaamphiphiles	
RFL_4FR	Nanosheets
RA_9R	Fibrils
KI_4K	Nanotubes
Ionic-complementary peptides	
EAK16, Ac-AEAEAKAKAEAEAKAK-Am	Fibrils
RADA16, Ac-RADARADARADARADA-Am	Fibrils

type I ionic-complementary peptides respectively (variants of these peptides corresponding to other types have also been studied). RADA16 was one of the first peptides developed for application in cell culture. This peptide and EAK16 form fibril-based gel scaffolds (in buffer, at sufficiently high concentration) which support cell growth.

3.12   CHARACTERIZATION METHODS FOR PEPTIDE ASSEMBLIES

3.12.1   Fluorescence and Dye Staining Methods

Fluorescence probe methods are widely used in the biosciences and as such have been utilized to investigate aspects of peptide aggregation, where this occurs for example due to the peptide having an intrinsic amphiphilic character, or having amphiphilic character resulting from attachment of lipid chains or polymers. Fluorescence probe assays may be used to determine the critical aggregation concentration (CAC), analogous to the critical micelle concentration (CMC) of surfactants. In addition, dye staining methods such as the use of Congo red staining can be used to detect amyloid structures. Molecular structures of fluorescent probe/dye molecules mentioned in this section are shown in Figure 3.23.

Pyrene is a fluorescent probe molecule which was developed to locate the CAC for conventional amphiphiles, as its fluorescence is sensitive to the local hydrophobic environment. It has more recently been used to detect aggregation due to hydrophobic effects in peptide systems. Pyrene fluorescence measurements of CAC usually involve determination of the ratio of third and first vibronic band intensities, I_1/I_3 at 373 nm and 393 nm, although sometimes I_1 itself shows discontinuities at the same concentration. Pyrene derivatives such as 1-pyrene carboxylic acid offer lower toxicity than pyrene itself and have also been successfully used in peptide CAC determination. The fluorescent probe ANS (8-anilino-1-naphthalenesulfonic acid, Figure 3.23) has also been used to locate CAC values. Like pyrene, ANS fluorescence is also sensitive to the local hydrophobic environment. It shows good fluorescence sensitivity and is less toxic than pyrene.

In contrast to use of pyrene or ANS, the fluorescence of thioflavin T (ThT) is dependent on the formation of amyloid-like structures and has been used for β-sheet fibril-forming peptides. Excitation of ThT at 450 nm produces fluorescence at 482 nm. ThT fluorescence has also been widely used to probe the kinetics of amyloid aggregation processes, as discussed further in Section 3.4.

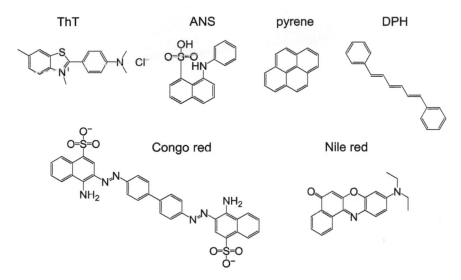

Figure 3.23 Molecular structure of fluorescent probe dyes mentioned in the text, used to stain peptide structures or in assays of aggregation. ThT denotes thioflavin T, ANS denotes 8-anilinonaphthalene-1-sulfonic acid, DPH is 1,6-diphenyl-1,3, 5-hexatriene.

Other fluorescent dyes such as Nile red (Figure 3.23) have been used to determine the CAC of amyloid-forming peptides. The fluorescent probe DPH (1,6-diphenyl-1,3,5-hexatriene, Figure 3.23) has also been used to locate the CAC of lipopeptides.

A further possibility for peptides containing tryptophan is to use the intrinsic tryptophan fluorescence to detect aggregation or unfolding. The intrinsic fluorescence of tryptophan can be used to probe the local environment of this residue since its emission peak is solvatochromic, i.e. it shifts to a lower wavelength in a more hydrophobic environment. Thus tryptophan fluorescence is used to examine protein folding and it can be used to locate the CAC for peptides that contain this residue.

Staining with fluorescent dyes is used in confocal microscopy and has been applied to image micron-scale self-assemblies. Congo red staining provides a method that identifies amyloid structure. Under polarized light, amyloid samples exhibit green birefringence when stained with Congo red. As mentioned in Section 3.3.1, it was used in 1906 by Alzheimer to detect brain tissue deposits. Figure 3.24 shows an example of a polarized optical microscopy (POM) image of a β-sheet forming lipopeptide stained with Congo red, showing the green birefringence texture. Förster resonance energy transfer (FRET) has been used to investigate the binding of dye to amyloid fibrils.

Figure 3.24 Congo red staining of β-sheet 'amyloid' fibrils formed by lipopeptide C_1–YEALRVANEVTLN. Source: From Castelletto et al. 2017.

3.12.2 Small-angle Scattering

Small-angle X-ray scattering (SAXS) and small-angle neutron scattering (SANS) are powerful complementary techniques to probe the nanostructure of self-assembled objects such as peptide assemblies. SAXS can be performed in the laboratory or at a central synchrotron facility, whereas SANS is performed using either a spallation or nuclear reactor as the source of neutrons. In either technique, the variation of the scattered intensity with $q = 4\pi \sin \theta/\lambda$ (where 2θ is the scattering angle and λ is the wavelength) provides information on the interaction between self-assembled objects via the structure factor, and also on the shape and scattering density profile (from the form factor). Since X-rays are scattered by electrons, a SAXS form factor depends on the electron density profile, whereas SANS provides information on the scattering length density profile, which depends on the nuclear scattering factors. In dilute solution, only form factor scattering is obtained and this is used to obtain the dimensions of the peptide assembly. SAXS offers the potential for fast time-resolved measurements and higher throughput measurements. SANS enables contrast variation measurements using H_2O/D_2O mixtures, and longer measurements without the risk of beam damage of samples (which would lead to ionization and molecular disassembly) and that can occur with synchrotron SAXS. Both techniques can reliably be employed for solutions with peptide concentrations down to around 0.1 wt% (1 mg ml^{-1}).

The SAXS form factor from unaggregated peptides (whatever the secondary structure, assuming the peptide is small) can readily be distinguished from that of peptide assemblies. The monomer form factor has a characteristic flat shape at low q, curving over at high q. The shape of this form factor can be contrasted with those of assembled structures that feature broad maxima and minima at higher q and characteristic slopes of the q-dependent intensity at lower q. Larger peptides and proteins have form factors which provide information on the shape envelope of the folded structure. These can be compared to crystal structures (which have much higher resolution of course) when available.

Further information on peptide samples that either exhibit the formation of lyotropic liquid crystal phases (e.g. nematic, hexagonal-packed cylinder, or lamellar) or which align under shear flow, can be obtained through SAS. At most synchrotron SAXS beamlines (and on many laboratory instruments) and SANS instruments the data is measured on a two-dimensional area detector and it is straightforward to check the original SAXS or SANS pattern (before reduction to one-dimensional intensity profiles) for sample alignment which is manifested in anisotropy in the SAS pattern. In some cases, spontaneous alignment of the sample is observed during flow into the sample cell (often a capillary).

3.12.3 Electron Microscopy and Scanning Probe Microscopy

Electron microscopy and scanning probe microscopy provide direct imaging of peptide assemblies. Atomic force microscopy (AFM), which is a scanning probe microscopy method, provides images of peptide structures deposited on flat surfaces. To date, the method has mainly been used to structures in solutions dried onto planar solid substrates such as mica or silicon. Tapping mode AFM (in which the AFM tip oscillates close to the sample surface) provides images of nanoscale variations in height due to topography across the surface. AFM can also be used to quantify the properties of amyloid fibrils including conformation and orientation. Measurement of fibre mechanical properties using peak force microscopy with an AFM instrument has now yielded values of the elastic modulus for many amyloid fibril systems.

There are two main types of electron microscopy. Conventional transmission electron microscopy (TEM) is performed on dried films on carbon grids, which are usually stained with heavy-metal-containing compounds to enhance contrast. Scanning electron microscopy (SEM) is used to image surfaces of pieces of a sample, most typically prepared by drying, although frozen or partially hydrated specimens can be imaged by cryo-SEM or environmental SEM (ESEM) respectively. Sample preparation in cryogenic-TEM (cryo-TEM) involve cooling the sample to $-187\,°C$ in liquid ethane to vitrify the aqueous phase. This technique avoids artefacts caused by slow drying. Cryo-TEM is considered to be the best imaging technique for amphiphilic peptide nanostructures. Figure 3.25 shows representative cryo-TEM images from different classes of peptide nanostructures.

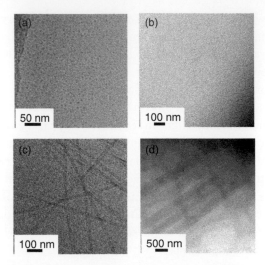

Figure 3.25 Representative cryo-TEM images from different classes of peptide nanostructures: (a) spherical micelles in a 0.5 wt% solution of C_{16}-CSK$_4$RGDS, (b), fibrils formed in a 0.5 wt% solution of lipopeptide PYY17 at pH 8, (c) twisted tapes formed in a 2 wt% solution of peptide GNNDESNISFKEK, (d) nanotubes formed in a 1 wt% solution of C_{16}-KKFFVLK.

SEM readily provides attractive images of peptide structures. However, SEM images should be interpreted with a degree of scepticism since a 'fibrillar network' morphology is observed for many peptide structures on the micron length scale that is not related to the individual nanostructure elements such as fibrils with 1–10 nm diameters, which can properly be resolved using (cryo-)TEM or AFM. Sample drying can cause artefacts in TEM and SEM imaging. AFM can be performed on sample surfaces under liquids, but this is a demanding imaging technique that is not routinely used.

It may be noted that in some cases, amphiphilic peptide assemblies such as fibrils or nanotapes are large enough (up to several μm in length) to be imaged by optical microscopy.

3.12.4   Analytical Ultracentrifugation

Analytical ultracentrifugation (AUC) involves centrifuging a peptide solution at very high rotation speeds (angular accelerations of tens of thousands of g, where g is the acceleration due to gravity) in order to determine the distribution of species, i.e. monomers and oligomers, present in a sample. This can be used to analyse the formation of coiled coils or small amyloid oligomers, for example. The sedimentation coefficient can be obtained and, with an estimate of the frictional ratio that accounts for the deviation from spherical shape of the particles, this can be used to obtain molar masses for the species present. AUC can be used for species with molar masses in the range from 100 Da up to megadaltons. The data is processed using software such as SEDFIT.

3.12.5   Gel Characterization Methods

The stiffness of peptide is gels is measured using rheological techniques. Rheology refers to the study of the flow properties of materials. Rheology methods are used to investigate the dynamic mechanical behaviour of peptide materials, in particular to measure the ridigity (modulus) of peptide gels. The simplest analysis of peptide gelation is inspection in an inverted tube. If the sample does not flow (over a reasonable timescale) it can be classified as a gel. This behaviour technically corresponds to that of a system with a finite yield stress.

The most common quantitative type of measurement to characterize peptide gels is to measure the dynamic shear moduli using a controlled stress or controlled strain shear rheometer. The gel is typically placed in a cone-and-plate or plate-plate tool in the rheometer and subjected to oscillatory shear or strain. The frequency-dependent measurements provide the dynamic storage (or elastic) shear modulus, G', and the dynamic loss (or viscous) shear modulus, G''. The measurements are performed in the linear viscoelastic regime, which is identified by first performing a stress or strain sweep to identify a stress or strain value within the range where the moduli (at a fixed frequency) do not depend on the stress or strain (typically the highest value in the range, to minimize noise on the measured data). Then frequency sweep measurements are performed at the selected low stress or strain value determined in the initial stress or strain sweep measurement. Figure 3.26 shows typical frequency-dependent modulus data for two samples of PAs. The PA C_{16}-GGGRGD forms a stiff hydrogel which is characterized by $G' > G''$ with both moduli largely independent of frequency. In contrast, the PA C_{16}-GGGRGDS does not form a hydrogel; the strong frequency dependence of the moduli and the fact that $G'' > G'$ are both properties that are characteristic of a liquid. Peptide gel moduli G' typically range from a few tens of Pa up to 10^5 Pa (at a frequency around 1 Hz). There is considerable scope to tune this by varying sample concentration, pH, buffer, and peptide sequence, etc.

Gels are sample-spanning networks. In the case of peptides, gels are often formed from entangled fibrils. The structure can be determined by *in situ*

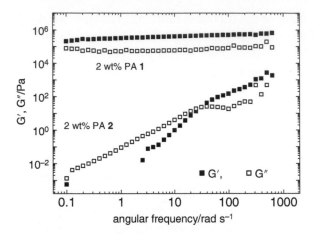

Figure 3.26 Dynamic shear moduli measured (at a stress σ = 100 Pa) for hydrogels of peptide amphiphiles PA **1** C$_{16}$-GGGRGD and PA **2** C$_{16}$-GGGRGDS. Source: From Castelletto et al. 2013.

SAS measurements. Imaging of gel morphology is typically carried out using SEM or cryo-SEM, bearing in mind the note of caution mentioned above about interpretation of SEM images. Samples for TEM, cryo-TEM, or AFM must be diluted in order to cover the grid or surface, and of course this disrupts the gel structure.

BIBLIOGRAPHY

Adamcik, J. and Mezzenga, R. (2011). Adjustable twisting periodic pitch of amyloid fibrils. *Soft Matter* 7: 5437–5443.

Adamcik, J. and Mezzenga, R. (2012). Study of amyloid fibrils via atomic force microscopy. *Current Opinion in Colloid & Interface Science* 17: 369–376.

Aggeli, A., Nyrkova, I.A., Bell, M. et al. (2001). Hierarchical self-assembly of chiral rod-like molecules as a model for peptide β-sheet tapes, ribbons, fibrils and fibres. *Proceedings of the National Academy of Sciences of the United States of America* 98: 11857–11862.

Arosio, P., Knowles, T.P.J., and Linse, S. (2015). On the lag phase in amyloid fibril formation. *Physical Chemistry Chemical Physics* 17: 7606–7618.

Badman, M.K. and Flier, J.S. (2005). The gut and energy balance: visceral allies in the obesity wars. *Science* 307: 1909–1914.

Blennow, K., Hampel, H., Weiner, M., and Zetterberg, H. (2010). Cerebrospinal fluid and plasma biomarkers in Alzheimer disease. *Nature Reviews Neurology* 6: 131–144.

Castelletto, V., Gouveia, R.J., Connon, C.J., and Hamley, I.W. (2013). New RGD- peptide amphiphile mixtures containing a negatively charged diluent. *Faraday Discussions* 166: 381–397.

Castelletto, V., Kaur, A., Kowalczyk, R.M. et al. (2017). Supramolecular hydrogel formation in a series of self-assembling lipopeptides with varying lipid chain length. *Biomacromolecules* 18: 2013–2023.

Caughey, B. and Lansbury, P.T. (2003). Protofibrils, pores, fibrils, and neurodegeneration: separating the responsible protein aggregates from the innocent bystanders. *Annual Review of Neuroscience* 26: 267–298.

Chiti, F. and Dobson, C.M. (2006). Protein misfolding, functional amyloid, and human disease. *Annual Review of Biochemistry* 75: 333–366.

Clemmensen, C., Muller, T.D., Woods, S.C. et al. (2017). Gut-brain cross-talk in metabolic control. *Cell* 168: 758–774.

Connon, C.J. and Hamley, I.W. (2014). *Hydrogels in Cell-based Therapies*. Cambridge Royal Society of Chemistry.

Das, A.K., Collins, R., and Ulijn, R.V. (2008). Exploiting enzymatic (reversed) hydrolysis in directed self-assembly of peptide nanostructures. *Small* 4: 279–287.

Dehsorkhi, A., Castelletto, V., and Hamley, I.W. (2014). Self-assembling amphiphilic peptides. *Journal of Peptide Science* 20: 453–467.

Fowler, D.M., Koulov, A.V., Balch, W.E., and Kelly, J.W. (2007). Functional amyloid – from bacteria to humans. *Trends in Biochemical Sciences* 32: 217–224.

Gazit, E. (2002). A possible role for π-stacking in the self-assembly of amyloid fibrils. *FASEB Journal* 16: 77–83.

Hamley, I.W. (2007). Peptide Fibrillisation. *Angewandte Chemie, International Edition in English* 46: 8128–8147.

Hamley, I.W. (2011). Self-assembly of amphiphilic peptides. *Soft Matter* 7: 4122–4138.

Hamley, I.W. (2012). The amyloid beta peptide: a chemist's perspective. Role in Alzheimer's and fibrillization. *Chemical Reviews* 112: 5147–5192.

Hamley, I.W. (2014). Peptide nanotubes. *Angewandte Chemie, International Edition in English* 53: 6866–6881.

Hamley, I.W. (2015). Lipopeptides: from self-assembly to bioactivity. *Chemical Communications* 51: 8574–8583.

Hamley, I.W. (2017). Small bioactive peptides for biomaterials design and therapeutics. *Chemical Reviews* 17: 14015–14041.

Henninot, A., Collins, J.C., and Nuss, J.M. (2018). The current state of peptide drug discovery: back to the future? *Journal of Medicinal Chemistry* 61: 1382–1414.

Hughes, M., Xu, H.X., Frederix, P. et al. (2011). Biocatalytic self-assembly of 2D peptide-based nanostructures. *Soft Matter* 7: 10032–10038.

Hutchinson, J.A., Burholt, S., and Hamley, I.W. (2017). Peptide hormones and lipopeptides: from self-assembly to therapeutic applications. *Journal of Peptide Science* 23: 82–94.

Hutton, J.C. and Siddle, K. (1990). *Peptide Hormone Secretion. A Practical Approach*. Oxford: Oxford University Press.

Ke, P.C., Zhou, R., and Serpell, L.C. et al. (2020). A half century of amyloid. *Chemical Society Reviews*, submitted.

Lau, J.L. and Dunn, M.K. (2018). Therapeutic peptides: historical perspectives, current development trends, and future directions. *Bioorganic & Medicinal Chemistry* 26: 2700–2707.

Linse, S. (2017). Monomer-dependent secondary nucleation in amyloid formation. *Biophysical Reviews* 9: 329–338.

Matson, J.B., Zha, R.H., and Stupp, S.I. (2011). Peptide self-assembly for crafting functional biological materials. *Current Opinion in Solid State & Materials Science* 15: 225–235.

Neal, J.M. (2016). *How the Endocrine System Works*. Chichester: Wiley-Blackwell.

Norman, A.W. and Henry, H.L. (2014). *Hormones*. Amsterdam: Academic Press.

Otzen, D.E. (2013). *Amyloid Fibrils and Prefibrillar Aggregates: Molecular and Biological Properties*. Weinheim: Wiley-VCH.

Papapostolou, D., Smith, A.M., Atkins, E.D.T. et al. (2007). Engineering nanoscale order into a designed protein fiber. *Proceedings of the National Academy of Sciences of the United States of America* 104: 10853–10858.

Shy, A.N., Kim, B.J., and Xu, B. (2019). Enzymatic noncovalent synthesis of supramolecular soft matter for biomedical applications. *Matter* 1: 1127–1147.

Tager, H.S. and Steiner, D.F. (1974). Peptide hormones. *Annual Review of Biochemistry* 43: 509–538.

Velichko, Y.S., Stupp, S.I., and Olvera de la Cruz, M. (2008). Molecular simulation study of peptide amphiphile self-assembly. *Journal of Physical Chemistry B* 112: 2326–2334.

Wang, H.M., Yang, Z.M., and Adams, D.J. (2012). Controlling peptide-based hydrogelation. *Materials Today* 15: 500–507.

Yu, M.Z., Benjamin, M.M., Srinivasan, S. et al. (2018). Battle of GLP-1 delivery technologies. *Advanced Drug Delivery Reviews* 130: 113–130.

4

Antimicrobial and Cell-penetrating Peptides

4.1 INTRODUCTION

Peptides with antimicrobial properties have evolved as part of the host defence mechanisms of many plants and organisms. More recently, novel antimicrobials have been designed based on such natural peptides, or from *de novo* design principles, based on the understanding of antimicrobial activity that has developed in recent years. This is discussed in this chapter. 'Antimicrobial' refers to activity against bacteria, fungi, viruses, or parasites. 'Antibiotic' is used as an alternative term, with the same general meaning. The main therapeutic applications of antimicrobial peptides are against bacterial and fungal infections. There is great interest in the development of new antibacterials in particular, due to the emergence of antimicrobial resistance (AMR) among many classes of bacteria, including multi-drug resistance (MDR). This is becoming increasingly problematic in the treatment of hospital infections and there is a lack of new treatments in the pipeline. However, the topic has recently attracted much attention and there are a number of promising candidate antibiotics undergoing advanced (phase III) clinical trials.

This chapter is organized as follows. Since most antimicrobial peptides (AMPs) are targeted towards bacteria, these are discussed first and in greater length. The first part of this chapter (Section 4.2) introduces the main types of bacterial pathogen, in particular the most important ones that have evolved antimicrobial resistance. The mechanisms by which antimicrobial agents can target bacteria, and the ways in which bacteria can develop

---

Introduction to Peptide Science, First Edition. Ian W. Hamley.
© 2020 John Wiley & Sons Ltd. Published 2020 by John Wiley & Sons Ltd.

resistance, are also summarized. Methods to test for antimicrobial activity are introduced in Section 4.3. Biofilms are surface coatings of bacteria that are particularly problematic in wounds, when covering medical devices, and in periodontitis and gum disease (for example gingivitis). The features of bacterial biofilms are considered in Section 4.4. Section 4.5 is devoted to an analysis of the design considerations for antimicrobial peptides. The primary classes of antibacterial peptides are presented in Section 4.6, along with selected exemplars of types of antibacterial peptides. Section 4.7 provides a short discussion of antifungal peptides, along with selected examples. Section 4.8 is concerned with antiviral peptides and Section 4.9 with peptides with antiparasitic activity. Then, in Section 4.10, various mechanisms that have been identified by which antimicrobial peptides interact with cell membranes are outlined. The chapter concludes with Section 4.11 on cell-penetrating peptides (CPPs). These share features in common with AMPs that interact with cell membranes. Cell-penetrating peptides are useful molecules which are being developed for gene therapies, for example, i.e. for delivering nucleic acids into cells.

4.2 BACTERIAL PATHOGENS, TARGETS OF ANTIBACTERIAL AGENTS, AND ANTIMICROBIAL RESISTANCE PATHWAYS

The leading cause of hospital-derived infections worldwide are the ESKAPE pathogens (*Enterococcus faecium, Staphylococcus aureus, Klebsiella pneumoniae, Acinetobacter baumannii, Pseudomonas aeruginosa,* and *Enterobacter* species). Table 4.1 lists pathogens that the World Health Organization (WHO) has identified as priorities for research into new antibiotics. Many of these species are commensals, species that mutually benefit from the presence of other species/strains. For example, there are harmless strains of *Escherichia coli, Enterococcus* spp., and *Klebsiella* spp. in the gut and of *S. aureus* in the nasopharynx (part of the upper respiratory tract) and some of these have evolved to be problematic due to emerging AMR.

Antimicrobial peptides (AMPs) are attractive since they can kill bacteria through non-specific mechanisms, for example lysis of the cell membrane. They can also act rapidly (within seconds), preventing the emergence of resistance. The mechanisms of activity of AMPs are discussed in more detail in Section 4.10.

AMPs are typically cationic and often contain arginine or lysine residues, although tryptophan-based AMPs are also known. Different classes of AMPs are discussed further in Section 4.6.

Table 4.1   List of pathogens identified by WHO as priorities due to antimicrobial resistance against existing treatments.

Priority	Pathogen	Resistance
Critical	*Acinetobacter baumannii*	Carbapenem-resistant
	Pseudomonas aeruginosa	Carbapenem-resistant
	Enterobacteriaceae[a]	Carbapenem-resistant, 3rd generation cephalosporin-resistant
High	*Enterococcus faecium*	Vancomycin-resistant
	Staphylococcus aureus	Methicillin-resistant, vancomycin intermediate and vancomycin-resistant
	Helicobacter pylori	Clarithromycin-resistant
	Campylobacter	Fluoroquinolone-resistant
	Salmonella spp.	Fluoroquinolone-resistant
	Neisseria gonorrhoeae	Third generation cephalosporin-resistant, fluoroquinolone-resistant
Medium	*Streptococcus pneumoniae*	Penicillin-non-susceptible
	Haemophilus influenzae	Ampicillin-resistant
	Shigella spp.	Fluoroquinolone-resistant

[a] *Enterobacteriaceae* include: *Klebsiella pneumonia*, *Escherichia coli*, *Enterobacter* spp., *Serratia* spp., *Proteus* spp., *Providencia* spp., and *Morganella* spp.; spp. indicates multiple species.

Antimicrobials have a range of distinct intracellular targets in bacteria, as shown schematically in Figure 4.1. The synthesis of bacterial nucleic acids is the target of quinolones such as ciprofloxacin; this compound disrupts DNA unwinding and replication via inhibition of DNA gyrase. AMPs in this class include indolicidin (Section 4.6.1) and PR-39 and buforin II (discussed in Section 4.6.2). RNA synthesis is prevented by molecules which block RNA polymerase, such as rifampicin (a polyketide) and the AMPs buforin II, human neutrophil peptide 1 (HNP-1) (see Section 4.6.1), and microcin J25 (from *E. coli*). Ribosomal syntheses of proteins in bacteria are the targets of several classes of non-peptide antibiotics (Figure 4.1), targeting the 30S or 50S bacterial ribosome subunit (these have sizes of 30 or 50 Svedbergs, a unit of sedimentation coefficient, see Section 3.12.4). A number of AMPs also target protein synthesis, including host defence peptides pleurocidin, PR-39, HNP-1, and magainin I. Magainin I also targets the cell wall and has antifungal activity (Section 4.7). Bacterial cell wall synthesis of peptidoglycans is the target of the well-known β-lactams penicillin, cephalosporin, carbapenems, and related compounds. AMPs with this type of activity include HNP-1 and the nisin peptides, which are discussed in more detail in Section 4.6.1. Synthesis of the essential vitamin folic acid,

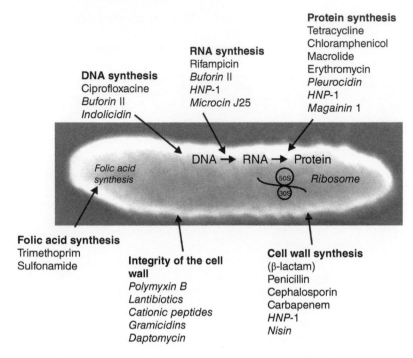

Protein synthesis
Tetracycline
Chloramphenicol
Macrolide
Erythromycin
Pleurocidin
HNP-1
Magainin 1

RNA synthesis
Rifampicin
Buforin II
HNP-1
*Microcin J*25

DNA synthesis
Ciprofloxacine
Buforin II
Indolicidin

DNA → RNA → Protein

*Folic acid
synthesis*

50S
30S
Ribosome

Folic acid synthesis
Trimethoprim
Sulfonamide

**Integrity of the cell
wall**
*Polymyxin B
Lantibiotics
Cationic peptides
Gramicidins
Daptomycin*

Cell wall synthesis
(β-lactam)
Penicillin
Cephalosporin
Carbapenem
HNP-1
Nisin

Figure 4.1 Targets for antimicrobials, showing examples for each type including examples of AMPs (*italicized*) and non-peptide antibacterial compounds in clinical use. The background image is a scanning electron microscope image of a *Pseudomonas* bacterium.

which occurs via dihydrofolate synthase in bacteria, is the target of the sulfonamides and trimethoprim. There are as yet no major AMPs for this target. Most peptide antimicrobials (including those listed in Figure 4.1, all of which are discussed in more detail later in this chapter) target the bacterial cell membrane, disrupting it via mechanisms discussed in Section 4.10. Other types of antimicrobials not shown in Figure 4.1 include bacterial protein-folding inhibitors (chaperone inhibitors) and inhibitors of proteases such as matrix metalloproteases (which are involved in production of extracellular matrix). In addition, several AMPs can inhibit bacterial cell division, including indolicidin and diptericin (an 83-residue peptide originally obtained from the blowfly *Phormia terranova*). Diptericin also has activity towards bacterial cell membranes. The diversity of activities of AMPs makes them attractive candidates for the development of new antimicrobial agents.

Antimicrobials may be targeted towards specific pathogens, or groups of pathogens such as Gram-negative or Gram-positive bacteria. The 'Gram' terms refer to a dye-staining method used to distinguish the two bacterial groups. These two types of bacteria are distinguished by differences in the

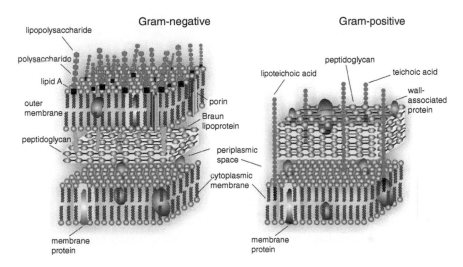

Figure 4.2   The distinct cell wall structures of the two different groups of bacteria.

composition of the bacterial cell wall (Figure 4.2). Gram-positive bacteria take up the Crystal violet stain used in the Gram test due to the presence of a thick outer peptidoglycan layer which is absent in Gram-negative bacteria. Gram-negative bacteria have an additional outer membrane which is coated in lipopolysaccharide (LPS) and which contains water-permeable channels formed by porin proteins. Well-known examples of Gram-negative bacteria include *E. coli* and *P. aeruginosa*, while Gram-positive bacteria include *S. aureus* and *Listeria monocytogenes*.

Fungi have a distinct typical cell wall structure, with the exterior cell wall comprising mannoproteins covering a layer of β-glucans that coat a chitin-coated lipid membrane. This is discussed further in Section 4.7.

The lipid membranes of many bacteria are rich in anionic cardiolipin (CL), zwitterionic phosphoethanolamine (PE), and/or anionic phosphoglycerol (PG) lipids, as discussed further in Section 4.10. This is in contrast to mammalian cell membranes, which are rich in zwitterionic phosphocholine (PC) lipids.

As well as the Gram stain classification of bacteria, it should be remembered that the morphology of bacteria can be very different. For example, *Staphylococci* and *Streptococci* bacteria are rounded whereas *Escherichia*, *Pseudomonas*, and many others are rod shaped. Rod-shaped bacteria are known as bacilli. Finally, it should be noted that some Gram-negative bacteria are surrounded by attachment pilus structures also known as fimbriae. These are amyloid structures, and the curli fibres discussed in Section 3.7 are one example of fimbriae. Fimbriae can be classed as a type

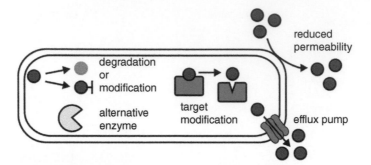

Figure 4.3   Antimicrobial resistance mechanisms.

of virulence factor; virulence factors are discussed further in Section 4.4. In addition, motile bacteria have attached flagella, which are motors that enable them to 'swim'.

Bacteria have or can evolve resistance mechanisms against antimicrobial agents, as shown in Figure 4.3. A bacterium can simply degrade or modify the AMP, modify the target (DNA or RNA), or it can use alternatives to the target enzymes in intracellular processes (or otherwise change the bio-chemical pathway that the antibiotic acts through). Bacteria can change the composition of the membrane to reduce permeability or they can increase the production of efflux pumps. Random genetic mutations can lead to AMR but resistance genes can also be transmitted from bacteria of one species, or across species, by horizontal gene transfer. Many of these mobile resistance genes encode enzymes that degrade the antimicrobial, for example β-lactamases that degrade β-lactam antibiotics.

Efflux pumps are transporters present in the membranes of Gram-negative bacteria that remove molecules, including antibiotics, from the cells. The production of efflux pumps can be up-regulated in biofilms. (Biofilms are attached coatings of bacteria which are discussed further in Section 4.4.) *Pseudomonas aeruginosa* shows enhanced efflux pump activity, for example, when the bacteria are growing in biofilm form.

Bacteria can also evade antimicrobials by altering the target. A good example of this is MRSA (methicillin-resistant *Staphylococcus aureus*), which has developed resistance to the β-lactam antibiotic methicillin by changing the cell wall synthesis enzymes from the one that is usually targeted to one that is not recognized by methicillin (or other β-lactams).

Another mechanism of AMR is reduced permeability of the membrane. This is achieved in Gram-negative bacteria by alterations in the composition of the outer lipopolysaccharide layer and/or by loss of porins in the outer membrane (Figure 4.2) or by mutations in their structure that restrict entry of the antimicrobial agent.

4.3 TESTING ANTIMICROBIAL ACTIVITY

Antimicrobial activity of different agents is often compared based on quantitative measures such as the minimum inhibitory concentration (MIC), which is the lowest concentration at which colony growth is visibly inhibited. The minimum bactericidal concentration (MBC) refers to the concentration of agent that kills the bacteria (typically by 99.9%). Since MBC is a more stringent measure, the MBC is usually a higher concentration than the MIC.

The simplest method to test antimicrobial activity for bacteria and fungi in planktonic form is based on colony counting of bacteria coating a Petri dish. Considering bacteria, in the most common methods the microbes are spread on the agar on the plate or suspended in the agar. The bacteria may need to be serially diluted in order to achieve a countable number on a plate. The number of colony-forming units (CFU) is then counted manually or using image processing software. For dense rapidly growing bacterial samples, turbidity measurements (turbidimetry) via light scattering can be an alternative; turbidimetry measures increases in the optical density as the bacteria grow. Usually CFU/ml of solution (or CFU/g of solid) is quoted in a dilution series and is commonly presented on a logarithmic scale. Alternatively, the disk diffusion test, also known as agar diffusion test, can be used; this involves measuring the zone of inhibition around a spot of the antimicrobial agent placed on an agar plate (Figure 4.4). Using either method with a range of concentrations of AMP enables determination of the MIC or MBC.

In vivo tests are required to develop antimicrobials for medical use; these may initially involve animal models before translation to human clinical trials. A simple invertebrate model that can be used in the early stage of screening of AMPs makes use of the larvae of the greater wax moth, *Galleria mellonella*. These are inexpensive, easy to handle (at 37 °C and having a short life span), and their use does not require ethical approval. The pigment of these larvae darkens (so-called melanization) in the presence of pathogens.

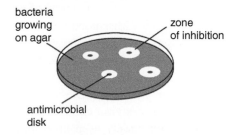

Figure 4.4 Measurement of inhibition zone of antimicrobial agent on a Petri dish.

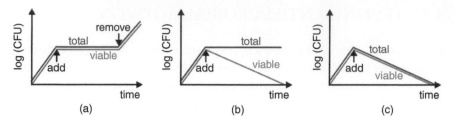

Figure 4.5 Different classes of antimicrobial agents. (a) Bacteriostatic agents inhibit growth but do not kill bacteria, growth resuming after the bacteriostatic agent is removed. (b) Bactericidal agents kill cells but do not lyse them. (c) Bacteriolytic agents kill bacteria by lysis.

Activity of AMPs against biofilms can be quantified by Crystal violet staining (this is the same dye as used in the Gram test). An additional test uses the dye Congo red (see Figure 3.23), which binds to lipopolysaccharides released by stressed bacteria.

Three distinct types of antimicrobial agent have been identified, according to whether they have bacteriostatic, bactericidal, or bacteriolytic activity. The distinctions are based on differences in the time-dependence of viable versus total cell count (expressed as number of CFU) as shown in Figure 4.5. Bacteriostatic agents hinder growth (by inhibiting biochemical processes) but as soon as they are removed, growth resumes. In contrast, bactericidal agents kill bacteria, but the dead cells are not lysed so the total cell count remains constant. Bacteriolytic agents kill bacteria and lyse them, so the numbers of both viable and total bacteria decline.

4.4 BACTERIAL BIOFILMS

The term for bacteria (or fungi) present as isolated cells in solution is planktonic. Many simple antimicrobial tests are based on analysis of planktonic bacteria. However, when bacteria spread they form biofilms, which are dense surface coatings of bacteria surrounded by a matrix of polysaccharides, proteins, and a variety of toxin species (discussed below). Curli fibres are produced by enterobacteria such as *E. coli* as part of the extracellular matrix within the biofilm. These are amyloid fibres formed by proteins secreted into the extracellular milieu (this is discussed further in Section 3.7).

Biofilms are responsible for many persistent infections and for contamination of medical devices (valves, catheters, contact lenses, etc.). *Pseudomonas* species have been widely used in biofilm studies since they are easy to culture and also because they are important human pathogens (and animal and plant microbes). *Pseudomonas aeruginosa* is responsible for hospital infections, including some types of sepsis and pneumonia. Several antimicrobial-resistant strains have also evolved and Gram-negative *P. aeruginosa* can thus be used as one model organism to study AMR. (Gram-negative *E. coli* and Gram-positive *S. aureus* are also used for this.) *P. aeruginosa* plays an important role in lung infections in those suffering from cystic fibrosis due to biofilm formation in the lung epithelium. In another example, biofilm formation is also important in cholera, when the *Vibrio cholerae* bacteria that cause cholera form colonies in the small intestine. It is thus an important challenge to develop AMPs that are active against biofilms.

Biofilms play an important role in AMR because: (i) the exposure to the antimicrobial agent is longer lasting, leading to more selection pressure of resistant microbes, (ii) sub-inhibitory concentrations resulting from reduced penetration of the antimicrobial can lead to specific resistance mechanisms, and (iii) horizontal gene transfer can be facilitated by the high cell density and increased genetic competence within biofilms. Biofilms can also harbour persister cells; these are cells that cannot grow in the presence of the antibiotic but which stay in a dormant state, ready to resume growth after removal of the antimicrobial. Persister cells are believed to be activated by stress responses of the bacteria. The presence of a population of persister cells can be inferred from the shape of the bacterial kill kinetic profile. The development of persistence can be contrasted with tolerance, in that in the former case only a sub-population survives the antimicrobial, whereas in the latter case there is reduced antimicrobial activity across the whole population. Tolerance can develop because of the slow growth rate in biofilms due to altered cell metabolism as well as the gradients in diffusion of the antimicrobial due to the extracellular matrix density and oxygen/nutrient gradients. In addition, tolerance can result from external stress factors (including starvation) and host factors (such as differences in host immunities), as well as the presence of antibiotics. Tolerance due to a slow growth rate can be inherited when a bacterial species or strain has an intrinsically low growth rate or can be non-inherited and result from the growth conditions. *Mycobacterium tuberculosis* is an example of a bacterium with an intrinsically slow growth rate. Dormancy can be viewed as the limiting case of zero growth rate. Tolerance can also develop during the lag phase of the bacterial growth phase, which is the delay period before growth-arrested cells resume growth. Tolerance can occur when the antibiotic treatment period is shorter than the timescale of the arrested growth.

The stages in the growth of a biofilm are shown schematically in Figure 4.6, along with a summary of strategies to prevent or interfere with

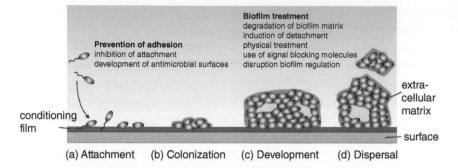

Figure 4.6 Stages in biofilm formation and strategies to prevent/disrupt or treat them, some of which can involve AMPs.

biofilm formation or development. The first stage of biofilm development occurs when bacteria initially attach to the surface. The surface may be covered by a conditioning film of already deposited organic material including proteins, polysaccharides, etc. Colonies then develop as the bacteria communicate and produce extracellular matrix. Further development of colonies then occurs, before the biofilm is able to disperse bacteria to further spread and form new colonies.

The antimicrobial killing kinetics differ in the presence of persister cells, in a population that has developed resistance, or in a population with antibiotic tolerance, as shown in Figure 4.7. The presence of persister cells leads to a characteristic bimodal (or multimodal) shape of the colony count decay curve. The number of colony-forming units actually increases in the case of bacteria that develop resistance. Tolerant bacteria are characterized by a slower killing rate than normal. Some populations of persisters may also have different lag phases and some may form in a dose-dependent manner.

It is far harder to develop effective antimicrobial agents against biofilms than against planktonic microbes because the biofilms restrict penetration of the antimicrobial compound, and because the growth rate of bacteria is lower since antimicrobials generally act on rapidly growing bacteria. The morphology of cells may also be distinct when attached to a surface within a biofilm. The expression of biofilm-specific resistance genes is also possible. In addition, the microenvironment within the biofilm may antagonize antimicrobial activity in regions where nutrients are depleted or where waste products have accumulated. MIC and MBC values quantifying activity against planktonic microbes often have no relation to activity against bacteria in biofilm form. Penetration of AMPs into biofilms can be monitored by fluorescence labelling of the peptide.

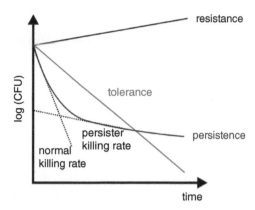

Figure 4.7 Antimicrobial killing kinetics for cultures showing resistance, tolerance, or the effect of persister cells.

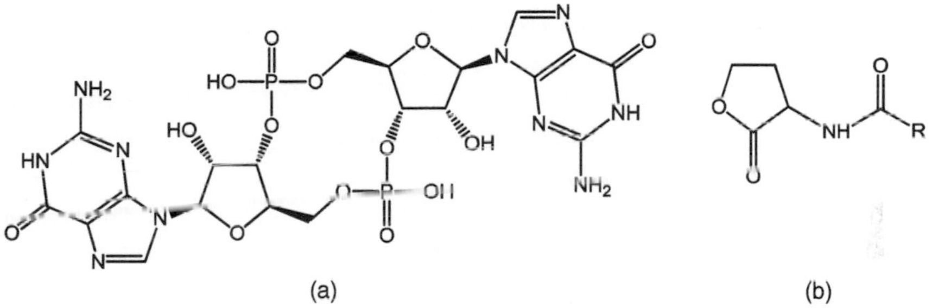

Figure 4.8 (a) Molecular structure of c-di-GMP. (b) Generic molecular structure of an acylated homoserine lactone (AHL) autoinducer.

Biofilm formation is triggered in many species by the cyclic nucleotide c-di-GMP (cyclic di-guanosine monophosphate) (Figure 4.8a). Disruption of biofilms is important in many clinical applications, for example in the treatment of infected wounds or to treat surfaces of medical devices. The process of quorum sensing plays an important role in bacterial biofilm formation as bacteria sense their local population density. This process involves gene regulation in response to the presence of a sufficient concentration of inter-cellular signalling molecules (autoinducers) within a growing bacterial population. Typical autoinducers for Gram-negative bacteria are acylated homoserine lactones (AHLs) (Figure 4.8b). Short peptides often serve as autoinducers for Gram-positive bacteria such as *S. aureus*. Blocking bacterial signalling molecules using peptides is a potential new strategy to use AMPs in treating biofilm infections.

Peptides also have potential in the targeting of virulence factors as a means to control infections. Virulence factors are produced by microbes and they enable the microbes to colonize a host or to enter host cells. Virulence factors also facilitate avoidance by the immune system and/or immunosuppression. Bacterial virulence factors include endotoxins, such as lipopolysaccharide (LPS) from the cell wall of bacteria, and various enzymes. Examples of enzyme virulence factors include hyaluronidase and collagenase, which act on the extracellular matrix materials hyaluronic acid and collagen respectively, and proteases and lipases, which act on lipid membranes. Exotoxins are toxic proteins such as haemolytic toxins, and are released by spreading pathogens. There are many other examples of virulence factors and bacterial biofilms contain a mixture of these toxins as well as polysaccharides, proteins, and the bacteria themselves.

4.5   DESIGN OF ANTIMICROBIAL PEPTIDES

Antimicrobial peptides (AMPs) may be designed based on known natural antibiotic peptide sequences or *de novo* based on known common components such as arginine or tryptophan residues (as discussed further in Section 4.6.2). Screening for new AMPs is often done by high-throughput combinatorial methods, using libraries of related sequences prepared by design or split-and-mix methods (see Section 2.5). These processes introduce sequence variability when batches of solid-supported peptides are split and mixed after each amino acid coupling, leading to a library with one unique peptide per starting bead. Other peptide library procedures may be employed. Structure–activity relationships may then be explored systematically. Quantitative structure–activity relationships (QSARs) can be established using the MIC or MBC to quantify the activity. A variety of descriptors relating to the peptide structure have been used, including the measured helicity, the Chou–Fasman α-helix propensity (see Figure 1.9), the Kyte–Doolittle hydrophobicity (see Table 1.1), the hydrophobic moment (see Eq. (1.5)), the net charge, the molecular weight, NMR parameters, the solvent-accessible surface area, and many others.

Computational screening of AMPs is also possible, based for example on assessment of selectivity of membrane interactions towards AMPs with compositions similar to those of bacteria (discussed further in Section 4.10). Molecular dynamics simulations of interacting peptide/lipid membranes are now possible at a sufficiently large scale to be able to elucidate modes of interaction at an atomistic level. Most recently, machine learning artificial intelligence principles have also been employed in an attempt to 'automatically' extract the relevant peptide features associated with useful activity.

Table 4.2   Marketed antimicrobial peptides.

Peptide	Structure	Mechanism of activity	Comments
Gramicidins	Gramicidin D is a mixture of linear 15-mer Gramicidins A, B, C. Cyclic peptide (gramicidin S)	Disrupt microbe cell walls (increases membrane permeability)	Used mainly against Gram-positive bacteria
Polymyxin B and colistin (polymxyin E)	Lipidated cyclic peptides	Disrupt microbe cell walls	Used as last resort antibiotics against infections caused by several Gram-negative bacteria
Daptomycin	Lipidated cyclic peptide	Disrupts microbe cell walls (pore formation)	Used to treat multi-drug-resistant Gram-positive bacterial infections
Valinomycin	Cyclic dodecadep-sipeptide derived from *Streptomyces fulvissimus*	Causes cell death via potassium ion transport across membranes	Highly toxic
Nisin	Polycyclic lantibiotic	Disrupts microbe cell walls (binding to lipid II)	Broad spectrum activity against Gram-positive bacteria. Used in food processing.

A number of peptide antimicrobials are on the market as therapeutics. These are listed in Table 4.2 and are discussed further in the following sections of this chapter. The term 'depsipeptide' (valinomycin in Table 4.2 is an example) refers to a peptide derivative in which ester groups replace amide groups (see Section 2.7). There is intense research to develop new AMPs and Table 4.2 will surely be extended as these proceed through clinical trials into practice.

To be practical in therapeutic applications, AMPs need to have (in addition to excellent antimicrobial activity, i.e. low MIC/MBC) low cytotoxicity towards human cells, high haemocompatibility, suitable pharmacokinetics, and reasonable cost.

For application in the treatment of infected wounds, an antimicrobial agent must be effective against bacteria in biofilms. Antimicrobial agents to treat surfaces, for example on medical devices or organs/biostructures (such as the lungs in cystic fibrosis or the teeth in dental caries), should prevent or hinder biofilm formation. For the former applications, peptides may be covalently attached to the surface or added as functional groups to polymer coatings. Effectiveness against biofilms may involve action against one of the biofilm-resistance mechanisms mentioned in Section 4.4, i.e. enhancing penetration of the AMP or stimulating antimicrobial activity under conditions of slow growth or chemical gradients. Peptides that interfere with quorum sensing or release of virulence factors have also been developed.

Due to issues of oral bioavailability of peptides, discussed in more detail in Section 5.2, peptide antimicrobials are generally delivered by other methods, in particular by parenteral administration (injection), by intranasal delivery, or topically. A few AMPs can be formulated for oral administration (e.g. polymyxins). On the other hand, nisin is incorporated into foods and food packaging during processing. Gramicidins are applied topically to treat wounds, as are many other antimicrobials. A number of strategies are available to improve the stability of peptides *in vivo* and these are also being investigated in the development of novel AMPs. Such strategies include attachment of lipid chains (lipidation), attachment of PEG (polyethylene glycol) polymer chains, or glycosylation. Lipidation and glycosylation are used in nature to increase serum stability, with these modifications being achieved post-translationally. These strategies are discussed further in Sections 2.8–2.10.

The activity of AMPs may also be improved in a combination therapy where an AMP with a given activity is combined with another AMP (or small molecule antibiotic) with complementary activity. Controlling the dosage is another broad area offering control over the management of AMR, for example cycling the use of different antibiotics over different periods or mixing the use of different antibiotics across a patient population (also at variable times).

4.6   CLASSES OF ANTIBACTERIAL PEPTIDES

This section covers a range of natural peptides, expressed by various organisms as part of their host defence system, as well as designed peptides. In terms of the source of natural AMPs, as of 2019 the AMP database available at http://aps.unmc.edu/AP lists 3142 AMPs, of which 74% are from animals, 11% from bacteria, 11% from plants, and the remainder from fungi, protists, and archaea. Among animal-derived peptides, the majority (more than 1000) are from amphibians (mostly frogs), with the next highest number being derived from mammals (including humans), while another major source is arthropods (insects and spiders, etc.) with more than 500 identified. These natural AMPs are characterized typically by a net charge in the range +1 to +6. They are typically 20–50 residues long and contain 30–70% hydrophobic residues. This percentage of hydrophobic residues is shifted notably towards the higher end for amphibian-derived peptides, which are also on average shorter and have a lower net positive charge. This database also reveals key chemical modifications in the natural AMPs, of which C-terminal amidation and cyclization (N-terminal to C-terminal or thioether bridges) are by far the most common types. In terms of secondary structures, the α-helix is predominant, except for the relatively small number of plant AMPs examined, for which β-sheet or mixed α-β structures were noted. Among animal-derived AMPs, arginine residues and lysine residues are present at abundances notably higher than those usually found in proteins, as expected given the typical net positive charge of AMPs.

As will become evident, the classes of AMPs discussed in the following are not mutually inclusive. For example, host defence peptides (HDPs) can be tryptophan-rich and/or rich in cationic residues (arginine/lysine).

4.6.1 Host Defence Peptides

There are many classes of host defence peptides (HDPs). Some examples are shown in Figure 4.9. The defensins are large peptides/small proteins found in many organisms, including both plants and animals. They are cationic and contain cysteine residues. They act by disrupting microbial cell membranes. Figure 4.9 shows a schematic molecular structure of one example (human neutrophil peptide 1, HNP-1, also known as α-defensin), showing the nested tricyclic structure that is stabilized by disulfide bonds between cysteine residues as indicated. Different classes of defensins (α-, β-, θ-, and others) have a characteristic pattern of cysteine residues and disulfide bridges. It has been proposed that some defensin peptides also form so-called 'nanonets', i.e. fibrillar mesh structures, that trap and surround bacteria or fungi, hindering biofilm formation. This provides an additional mode of action.

The cathelicidins are AMPs that are important in the mammalian innate immune system. Most cathelicidins are linear peptides with α-helical conformations. An important example of a human cathelicidin of great interest as a basis for the development of new antimicrobials is LL-37 (Figure 4.9). Many variants of LL-37 based on truncations and/or modifications have been examined as antimicrobial agents. LL-37 is also over-expressed in

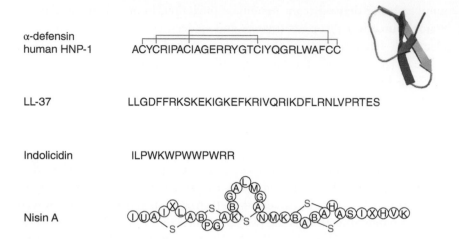

α-defensin
human HNP-1   ACYCRIPACIAGERRYGTCIYQGRLWAFCC

LL-37   LLGDFFRKSKEKIGKEFKRIVQRIKDFLRNLVPRTES

Indolicidin   ILPWKWPWWPWRR

Nisin A

Figure 4.9 Molecular structures of some host defence AMPs: (a) human neutrophil peptide 1 (HNP-1, also known as human α-defensin 1) showing three-dimensional pdb structure (right) with sequence with disulfide bond locations in red, (b) LL-37, (c) indolicidin, (d) nisin A in schematic form with thioether bridges shown (U = Dhb, 2,3-dehydro aminobutyric acid; X = Dha, dehydroalanine; B = Abu, aminobutyric acid).

chronic inflammatory diseases, including Crohn's disease, eczema, and rheumatoid arthritis. PR-39, mentioned in Section 4.2, is another catheli-cidin. Cathelicidins are believed to act via disruption of the cell membranes of microbes.

The bacteriocins are a large class of AMPs expressed by bacteria to fight infections caused by closely related strains of bacteria. The lantibiotics are a subgroup of bacteriocins that comprise a large number of AMPs containing lanthionine (see Table 1.3). They are produced by Gram-positive bacteria as part of their defence against other Gram-positive bacteria. There is a series of nisin lantibiotic peptides, including nisin A shown in Figure 4.9. Nisin A is used as a food preservative to inhibit Gram-positive bacteria in cheese and other food products. The other bacteriocins are termed non-lantibiotics.

Many organisms, including plants and animals, express natural AMPs. There are many databases that list these peptide sequences, along with their properties. Some are included in Table 1.6.

4.6.2   Arginine- and Tryptophan-based Peptides

Cationic peptides are the most common type of antimicrobial peptide (AMP). Arginine-rich cationic AMPs are widely found in nature and are used in designed sequences, although there are also many examples of lysine-based cationic AMPs. Arginine-based peptides are considered to have cell-penetrating properties (which may be useful for other applications; to deliver cargo, for example). In general their activity is due to disruption of microbe cell membranes resulting from interactions with anionic groups in lipopolysaccharides (in Gram-negative bacteria) and lipoteichoic acid/teichoic acids (in the case of Gram-positive bacteria) and/or with the anionic head groups of the membrane phospholipids. Arginine forms bidendate hydrogen bonds with phosphate or carboxylate groups (Figure 4.10).

Tryptophan is a large hydrophobic residue; however, its main role in AMPs is believed to arise from the π-electron system within the indole group, which gives rise to a quadrupole moment and enables cation–π interactions with cationic groups. An example of a tryptophan-rich AMP is indolicidin, which is a short host defence peptide (HDP) obtained from bovine neutrophils (Figure 4.9). It acts by interacting with Gram-negative bacterial membrane lipopolysaccharides. Examples of other arginine- and tryptophan-rich AMPs are shown in Figure 4.11.

The gramicidins are a series of tryptophan-rich 15-residue peptides active against Gram-positive bacteria, being derived from the soil bacterial species *Bacillus brevis*. The peptides have the sequence formyl-XGAlAvVvWlYlWlW-ethanolamine (with alternating L- and D- residues), where X is V or I and Y is W (gramicidin A, Figure 4.11a), F (gramicidin B), or Y (gramicidin C). Gramicidin D is a mixture of gramicidin peptides that is marketed as Neosporin/Sofradex for topical application against most Gram-positive and some Gram-negative bacteria. Tyrothricin is a mixture of gramicidin and tyrocidine, which is a mixture of cyclic decapeptides. Tyrothricin was one of the first antibiotics developed and is administered

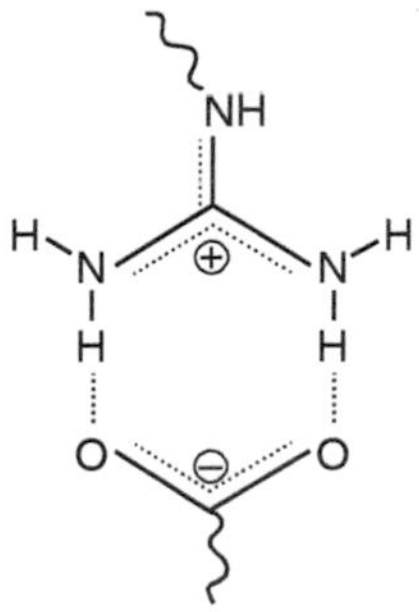

Figure 4.10 Bidentate binding of guanidinium groups with acidic XOO⁻ groups (X = C, P).

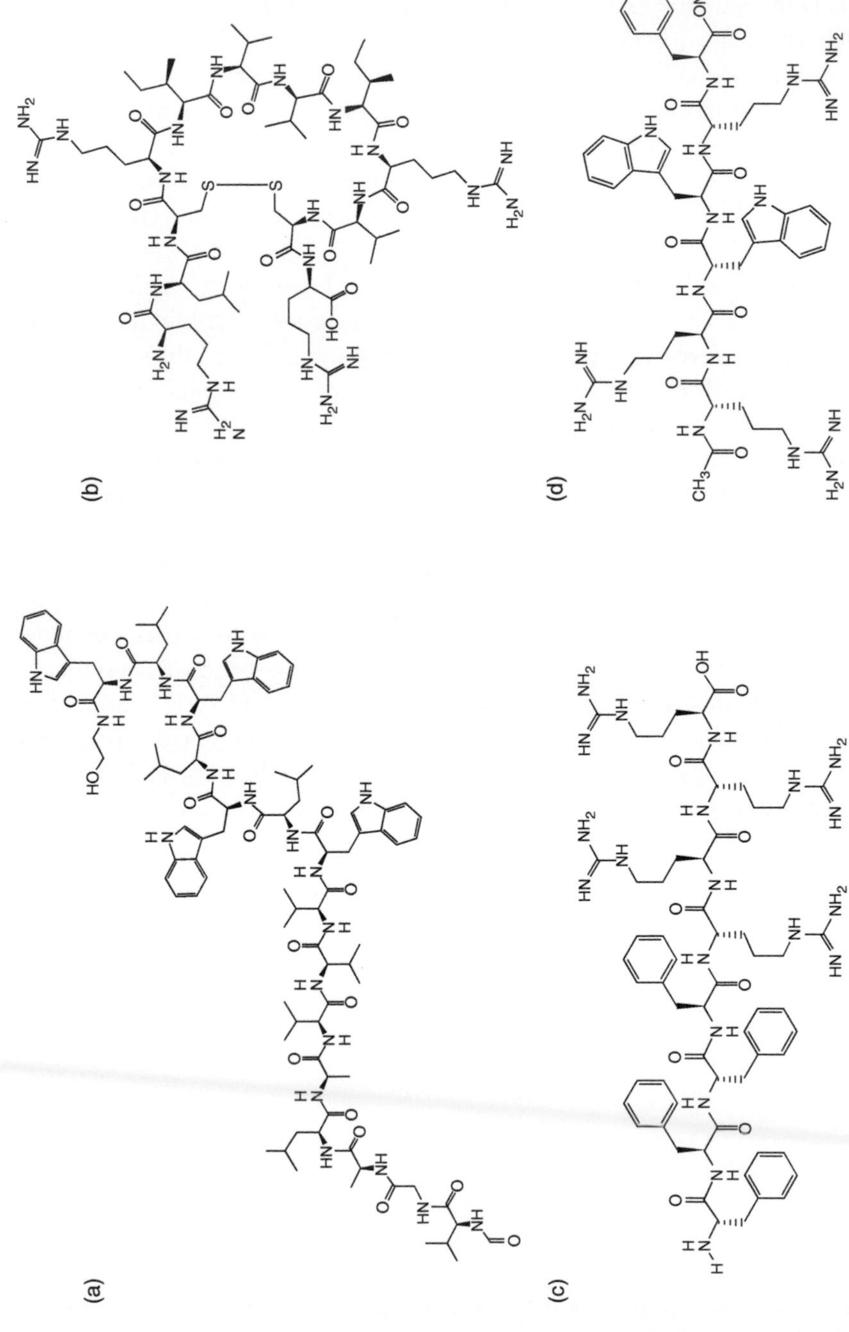

Figure 4.11 Molecular structures of some arginine-rich AMPs: (a) gramicidin A, (b) bactenecin 1, (c) model antimicrobial peptide R_4F_4, (d) model antimicrobial peptide RRWWRF.

orally. Gramicidin S is the cyclic decapeptide cyclo(VOLfP)$_2$ [O: ornithine]. Gramicidins are interesting AMPs without cationic residues. They form β-helix dimers that span the lipid bilayer, causing the formation of selective ion channels, rather than causing bacterial cell lysis due to electrostatic binding as in the case of the cationic AMPs discussed above. Bactenicin 1, shown in Figure 4.8b, is an example of an arginine-rich AMP. It is obtained from bovine neutrophils, has a cyclic architecture, and is one of the shortest bio-derived cationic AMPs. Examples of linear arginine-rich AMPs include buforin II (TRSSRAGLQFPVGRVHRLLRK), which is an amphibian HDP, drosocin, a 19-residue peptide from the fruit fly, and PR-39, a pig-derived peptide with a 39-residue sequence rich in R and P residues. Other examples of cyclic AMPs are discussed in Section 4.6.3. Figure 4.9c,d show examples of short designed AMPs rich in R and/or W residues.

4.6.3   Cyclic Peptides and Lipopeptides

Some examples of cyclic peptides and lipopeptide AMPs are shown in Figure 4.12. Daptomycin is a cyclic lipopeptide produced by the Gram-positive bacterium *Streptomyces roseosporous*. It is an example of an anionic AMP. These are expressed by many prokaryotes as part of their host defence system, and other examples are discussed in Section 4.6.4. Daptomycin is used as an antibiotic to treat serious infections caused by Gram-positive bacteria. It is effective against infections where AMR is a problem, including methicillin-resistant *Staphylococcus aureus* (MRSA) and vancomycin-resistant enterococci. As shown in Figure 4.12a, daptomycin is composed of 13 amino acids, 10 making up the cyclic structure and the other 3 forming a chain. The cyclic section of the molecule is linked through an ester bond to the tail through the terminal kynurenine residue (see Table 1.3, this is an unusual non-canonical amino acid product of tryptophan metabolism), and the threonine hydroxyl group. The bacitracins are a series of related cyclic lipopeptides with activity against Gram-positive bacteria due to their influence on peptidoglycan synthesis in the bacterial cell wall. One example is shown in Figure 4.12b. These compounds were originally derived from *Bacillus subtilis* and *Bacillus lichenformis*.

Polymyxin B lipopeptides (Figure 4.12c) are antimicrobial molecules produced by *Bacillus polymyxa*. Polymyxin B and polymyxin E (colistin) show antibiotic activity against a range of Gram-negative bacteria. The two molecules are distinguished by the substitution of a D-leucine in colistin with a D-phenylalanine in polymyxin B. The mode of action of these lipopeptides has been proposed to be membrane disruption due

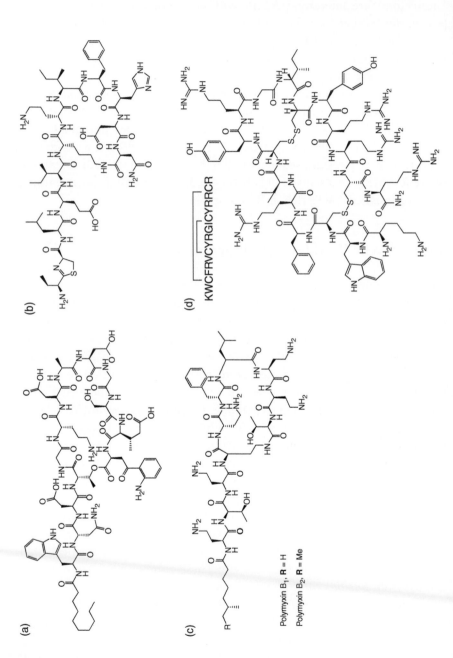

Figure 4.12 Molecular structures of some AMPs based on cyclic peptides: (a) daptomycin, (b) bacitracin A$_1$, (c) polymyxins B$_1$ and B$_2$, (d) tachyplesin 1, including molecular structure and schematic highlighting location of disulfide bridges.

KWCFRVCYRGICYRRCR

Polymyxin B$_1$, **R** = H
Polymyxin B$_2$, **R** = Me

to interaction between the cationic polymyxin and the anionic bacterial outer membrane leading to a detergent-like activity. Figure 4.12d shows an example of a bridged (bicyclic) AMP, tachyplesin A_1. The two tachyplesin AMPs contain four cysteines forming two disulfide bridges to give a bicyclic configuration. These peptides, originally obtained from horseshoe crabs, are highly charged, containing five cationic residues.

A wide range of antibiotics based on other cyclic depsipeptides (in which one or more amide groups are replaced by the corresponding ester group) or N-methylated peptides (peptoids) have been investigated.

4.6.4   Anionic Antimicrobial Peptides

Anionic antimicrobial peptides (AAMPs) tend to be long linear sequences, as exemplified by representative sequences shown in Table 4.3. Daptomycin, discussed in Section 4.6.3, is a counter-example. It may seem surprising that anionic peptides can interact with the anionic membranes of bacteria; however, this interaction is believed to be mediated via Zn^{2+} ions which coordinate with the anionic D and E residues. A number of AAMPs are present in the respiratory tract as part of the innate host defence system. These molecules are rich in aspartic acid (D) residues and are termed surfactant-associated AMPs. They are active against both Gram-negative and Gram-positive species. AAMPs are now known to also be a component of the innate immune system of the brain, and this along, with the blood–brain barrier and the meninges, provides protection against infections. Examples of human brain AAMPs are listed in Table 4.3. It has been proposed that the amyloid peptides Aβ40 and Aβ42 (see Chapter 3) have antimicrobial activities in the brain. Both these peptides are anionic at physiological pH, although the interaction with bacterial lipid membranes is believed to be via the cationic residues present in the sequences. AAMPs are also present in the epidermis as part of the skin defence barrier. Dermcidin is an example of one of these peptides. Dermcidin is also present in the brain. AAMPs are also present in blood plasma, for example fragments of the blood coagulant fibrinogen have AAMP properties. AAMPs may also be released by protease degradation of larger proteins in the gut. Fragments of milk proteins, including casein, α-lactalbumin, and β-lactoglobulin, have been identified that are rich in anionic residues and may have a role in killing microbes during digestion.

Table 4.3 Examples of anionic antimicrobial peptides.

Peptide	Sequence	Organs
Sequence identified from antibody studies	DDDDDDD	Lungs
Enkelytin from proenkephalin A	FAEPLPSEEEGESYSKEVPEMEKRYGGFM	Brain
Thymosin β_4	MSDKPDMAEIEKFDKSKLKKTETQEKNPLPSKETIEQEKQAGES	Brain
Dermcidin	SSLLEKGLDGAKKAVGGLGKLGKDAVEDLESVGKGAVHDVKDVLDSV	Epidermis/brain
Fibrinogen α-chain	DSGEGDFLAEGGGVRGPRVVERHQSACKDS	Blood plasma

4.7   ANTIFUNGAL PEPTIDES

The cell membrane of fungi is shown schematically in Figure 4.13; it differs from the cell membrane of bacteria because the lipid membrane is covered with chitin, which provides structural rigidity and which is itself coated with β-glucans with a final, outer surface comprising mannoproteins. The fungal cell plasma membrane is enriched in ergosterol, as distinct from cholesterol found in animal cell membranes. Ergosterols are therefore targets for antifungal agents. (Ergosterols are also found in protozoa and hence are targets for parasitic diseases.) The dominant fungal cell wall component is β-(1 → 3)-glucan, branched via β-(1 → 6) linkages. The outer mannoprotein layer provides structural and chemical protection to the fungal cell. Mannoproteins are glycoproteins rich in the saccharide mannose.

Table 4.4 lists selected peptides with antifungal properties. Examples of these peptides derived from animals are cyclic peptides stabilized by disulfide bridges. Antifungal peptides derived from insects and amphibians include linear α-helical peptides such as cecropin or the dermaseptins. Many of these also have antibacterial activity, for example the defensins are discussed in Section 4.6.1. Cyclic antifungal lipopeptide structures are exemplified by those shown in Figures 4.14 and 4.15.

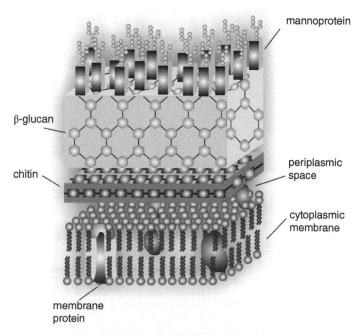

Figure 4.13   Schematic of the cell membrane of fungi.

Table 4.4 Examples of antifungal peptides.

Peptide	Source	Sequence[a]	Typical target fungus species
Defensin NP-1	Rabbit	$VVC_1AC_2RRALC_3LPRERRAGFC_3RIRGRIHPLC_2C_1RR$	Candida neoformans
Protegrin 1	Pig	$GGRLC_1YC_2RRRFC_2VC_1VGR$	Candida albicans
Lactoferricin B	Human	$FKC_1RRWQWRMKKLGAPSITC_1VRRAF$	Candida albicans
Cecropin B	Hyalophora cecropia (moth)	KWKVFKKIEKMGRNIRNGIVKAGPAIAVLGEAKAL	Aspergillus fumigatus
Dermaseptin S1	Phyllomedusa sauvagii (frog)	ALWKTMLKKLGTMALHAGKAALGAAADTISQGTQ	Candida neoformans
Magainin 2	Xenopus laevis (frog)	GIGKFLHSAKKFGKAFVGEIMNS	Candida albicans

[a]The subscripts indicate disulfide bridges between cysteine residues.

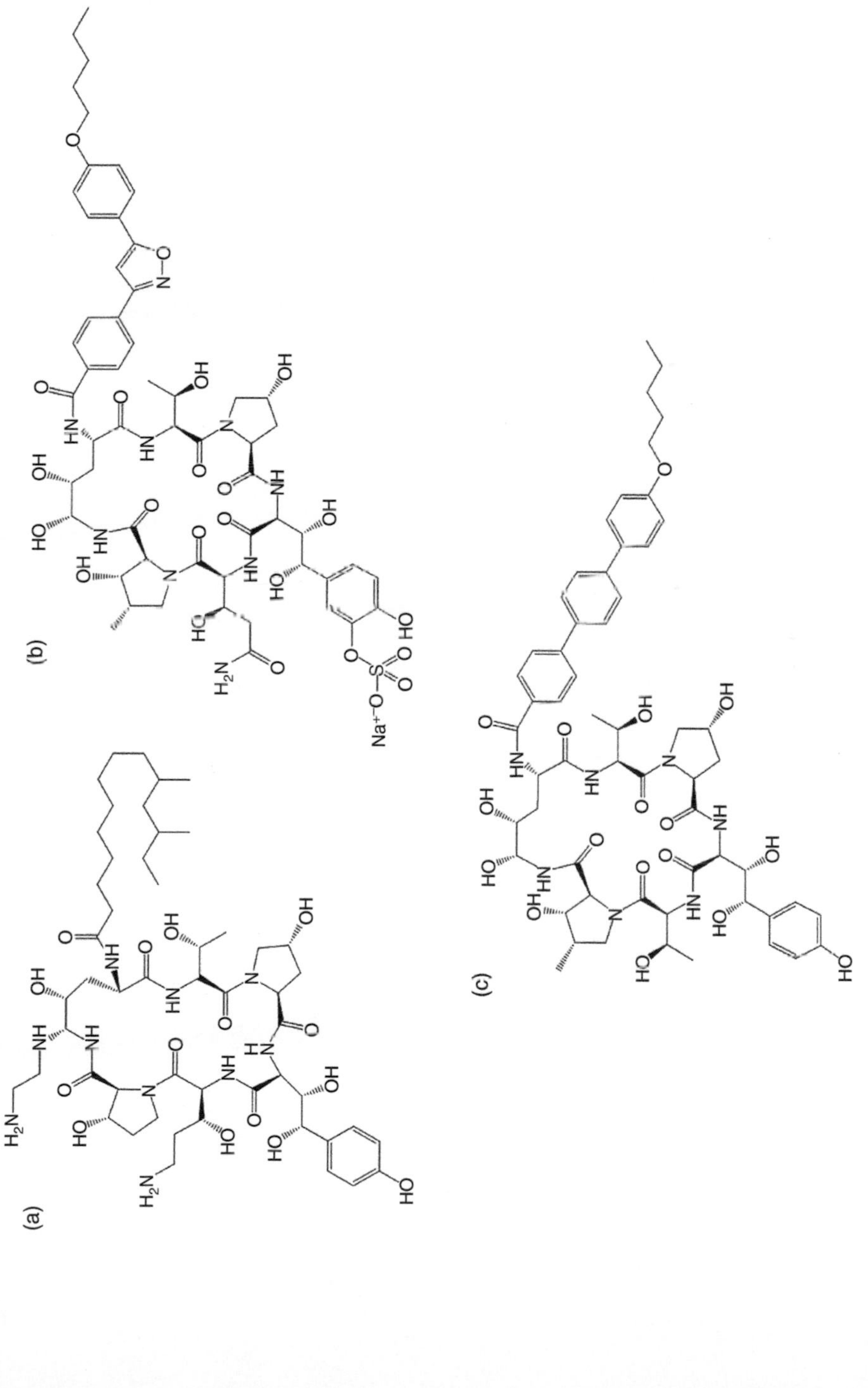

Figure 4.14 Molecular structure of several echinocandins: (a) caspofungin, (b) micafungin, (c) anidulafungin.

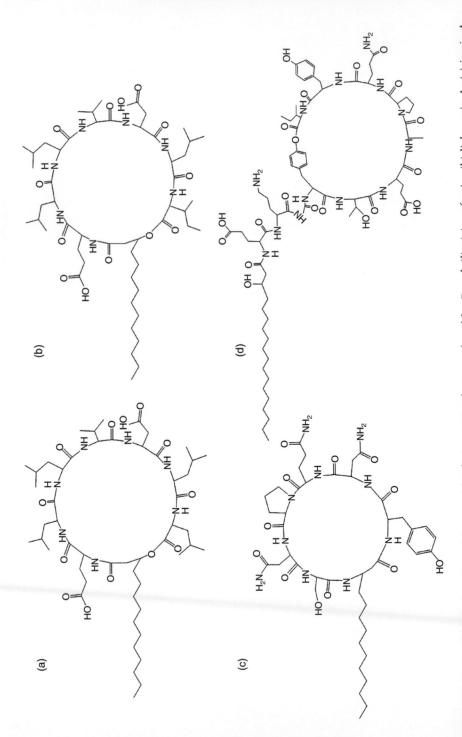

Figure 4.15 Examples of antifungal lipopeptides from among those produced by *B. subtilis*: (a) surfactin, (b) lichenysin A, (c) iturin A, (d) fengycin/plipastatin A.

The echinocandins, examples of which are shown in Figure 4.14, are a class of lipopeptide-based antifungal agents that are used clinically to treat yeast and *Candida* fungal infections, for example Caspofungin is marketed by Merck and Co. The echinocandins disrupt fungal cell glucan synthesis by inhibiting β-$(1 \rightarrow 3)$-glucan synthase enzyme activity.

Another class of antifungal agent are lipopeptides, such as those shown in Figure 4.15, which are expressed by *B. subtilis* as part of its host defence system. *Bacillus subtilis* mainly produces four families of lipopeptide biosurfactant: the surfactins, the lichenysins (with very similar structures to surfactins), the fengycins (or plipastatins), and the iturins (including mycosubtilin). These lipopeptides have antibacterial as well as antifungal properties and as the name suggests surfactins (and the other *B. subtilis* lipopeptides) have surfactant properties. These compounds are also being researched for other potential biomedical applications since some also show antiviral, antiparasitic, haemolytic, and anti-cancer activities.

Cyclic lipopeptides have also been obtained from *Pseudomonas* species, including members of the viscosin, amphisin, tolaasin, and syringomycin families. These typically have C_8–C_{12} lipid chains. The latter two families contain peptides with many non-natural residues (shown in Table 1.3) including diaminobutyric acid (Dab), 2,3-dehydroaminobutyric acid (Dhb), homoserine (Hse), *allo*threonine, and (in the syringomycin family) 4-chlorothreonine (4 ClThr). These lipopeptides have antimicrobial and biosurfactant activities.

The mechanisms of activity of antifungal peptides include the disruption of fungal cell glucan synthesis, as mentioned above, but in addition mechanisms of cell membrane disruption similar to those for antibacterial peptides (discussed in detail in Section 4.10) have been proposed, including lysis and pore formation. In addition to biosynthesis of glucan, the biosynthesis of chitin may be another mode of activity to be targeted.

4.8 ANTIVIRAL PEPTIDES

A number of peptides have been discovered to have antiviral activity against a range of viral infections, including HIV (human immunodeficiency virus), hepatitis C and B viruses, influenza virus, and herpes simplex virus (HSV). Antiviral peptides act by binding to heparan sulfate, a glycosaminoglycan (GAG) that facilitates binding to host cell membranes. The interaction of viruses with cell GAGs, which are anionic polysaccharides, can be blocked using cationic peptides. The presence of cationic residues (especially arginine) in antiviral peptides is similar to the cases of several types of antibacterial peptides (Section 4.6) and cell-penetrating peptides (Section 4.11). Antiviral peptides can also interfere with specific viral receptors on the host cell and they can block viral entry by interacting with viral glycoproteins. They can also interact with the viral envelope, which is a lipid membrane derived from host cell membranes that surrounds the protein capsid shell in some bacteria, along with viral glycoproteins. Finally, antiviral peptides can interact with host cell membranes (through which viruses enter cells) via translocation, pore formation, or lysis, similar to antibacterial peptides. These mechanisms are discussed in more detail in Section 4.10.

Table 4.5 lists examples of antiviral peptides. Other examples include indolicidin and LL-37, discussed in Section 4.6.1, the defensins, discussed in Sections 4.6.1 and 4.7, the dermaseptins, discussed in Section 4.6, the tachyplesins, discussed in Section 4.6.3, and the protegrins and magainins, discussed in Section 4.7. Polyphemusin has a sequence related to that of tachyplesin 1 and gomesin (another natural antibacterial peptide). Melittin (Table 4.5) is a cell-penetrating peptide.

4.9   ANTIPARASITIC PEPTIDES

Relatively little research has been done as yet on peptides with antiparasitic properties; however, a number of peptides have been shown to have activity against protozoa. Protozoa are the cause of many tropical diseases, such as leishmaniasis, malaria, and trypanosomiasis (African sleeping sickness). Table 4.6 lists examples of peptides with antiparasitic activity. Many of these peptides show antimicrobial and/or antifungal and/or antiviral activity and are also shown in Figure 4.9 or are listed in Tables 4.4 and 4.5, so the sequences are not repeated in Table 4.6.

Many of these peptides target the cell membranes of the parasites according to mechanisms similar to those for AMPs, as discussed in Section 4.10.

4.10   MECHANISMS OF ACTIVITY

4.10.1   Planktonic form

Modes of action of AMPs generally involve disruption of microbial membranes; however, see Figure 4.1 for other targets. In the following, we consider interactions with bacterial membranes although, as mentioned above, some antifungal peptides interact with their respective fungal

Table 4.5 Examples of antiviral peptides.

Peptide	Source	Sequence	Virus
Cecroprin A	*Hyalophora cecropia* (moth)	KWKLFKKIEKVGQNIRDGIIKAGPAVAVVGQATQIAK	Junin virus HSV
Melittin	*Apis mellifera* (honeybee)	GIGAVLKVLTTGLPALISWIKRKRQQ	HSV Junin virus
Brevinin-1	*Rana brevipoda porsa* (frog)	FLPVLAGIAAKVVPALFCKITKKC	HSV
Polyphemusin[a]	*Limulidae* (horseshoe crabs)	RRWC$_1$FRVC$_2$YKGFC$_2$YRKC$_1$R	HIV

[a]Subscripts indicate disulfide bridges between cysteine residues.

Table 4.6 Examples of antiparasitic peptides.

Peptide	Source	Activity against[a]
Cecroprin A	Moth	*L. donovani*
Magainin II	Frog	*P. caudatum, P. falciparum, C. parvum, Tetrahymena pyriformis*
Melittin	Honeybee	*L. donovani*
Buforin II	Frog	*C. parvum*
Temporin A	Frog	*L. donovani, L. mexicana*
Dermaseptin O1	Frog	*T. cruzi*
Indolicidin	Cow	*L. donovani*
HNP-1	Human	*T. cruzi*
Tachyplesin-1	Horseshoe crab	*L. braziliensis, T. cruzi*
Histatin-5	Mammals	*L. donovani, L. pifanoi*

[a]*L.* denotes *Leishmania, P.* denotes *Plasmodium, T.* denotes *Trypanosoma, C.* denotes *Crytosporidium.*

membranes (and antiviral peptides interact with host cell membranes) via similar mechanisms.

Figure 4.16 shows the three main proposed mechanisms. Typically these are based on interactions with the lipid membrane component of the membranes, and the outer polysaccharide/teichoic acid part is neglected. In the barrel stave mechanism, peptides aggregate and then insert into the lipid membrane with their hydrophilic faces forming the interior of the pores. The second model involves the insertion of the peptides into the membrane to form toroidal pores, in which the lipid membrane curves to accommodate the peptide molecules. In the carpet model, peptides lie on the membrane to form a carpet with subsequent detergent-like action to break up the membrane.

Membrane insertion by α-helical peptides often occurs with the peptide tilted with respect to the bilayer normal. This creates negative curvature in the membrane, promoting disturbance of the lipid packing. This can be detected by CD methods on aligned films and is also computationally modelled. Insertion of helical peptides into membranes is analysed in terms of tilt angles, which can be measured from NMR or CD on aligned samples of peptides in membranes.

When considering the activity of AMPs, it should be kept in mind that different bacterial membranes have distinct lipid compositions. Table 4.7 gives representative values. Lipid membranes or vesicles comprising mixtures of these lipids are often used as model systems, in comparison with phosphocholine (PC)-based lipids as models for mammalian cell membranes. Membrane-active AMPs should have selectivity against bacterial

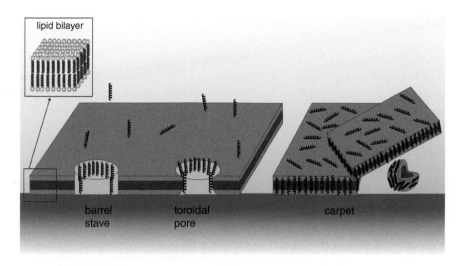

Figure 4.16 Membrane-disruption modes of antimicrobial activity. The peptide is represented as a α-helix. There are two kinds of membrane pore models – barrel stave and toroidal pore. In the carpet model, the membrane is broken up by peptide detergent action, forming membrane pieces and micelles.

Table 4.7 Lipid composition of the cell membrane of selected bacteria.

Species	% CL	% PG	% PE
Gram-negative bacteria			
Escherichia coli	5	15	80
Enterobacter cloacae	3	21	74
Klebsiella pneumoniae	6	5	82
Pseudomonas aeruginosa	11	21	60
Gram-positive bacteria			
Staphylococcus aureus	42	58	0
Streptococcus pneumonia	50	50	0
Bacillus polymyxa	8	3	60

Key: CL is cardiolipin, PG is phosphoglycerol (phosphatidylglycerol), PE is phospho-ethanolamine (phosphatidylethanolamine).

membranes compared to mammalian cells and in model systems lipid mixtures such as those in Table 4.7 may be used. However, this is a gross simplification, due to the presence of lipopolysaccharides (Gram-negative) or peptidoglycan/(lipo)teichoic acids (Gram-positive) at the exterior of the bacterial membrane (Figure 4.2) or mannoproteins at the surface of fungal cell walls (Figure 4.13).

4.10.2 Biofilms

As mentioned above, biofilms comprise bacteria embedded in a dense covering within extracellular matrix that is rich in polysaccharides. AMPs can act against biofilms by killing individual bacteria according to the mechanisms discussed above. In addition, amphiphilic AMPs can disrupt biofilms by detergent action, displacing biofilm molecules and suspending them in micelles. AMPs can also disrupt bacterial signalling molecules and interfere with quorum sensing (Section 4.4). The spreading of bacteria can also be hindered when they become trapped within AMP meshes (termed 'nanonets'), as mentioned in Section 4.6.1.

4.11 CELL-PENETRATING PEPTIDES

Cell-penetrating peptides (CPPs) and related protein transduction domains (PTDs) are of great interest since they have been shown to be able to transport drug molecules into cells. The drug molecule may be covalently bound to the CPP or it may be formulated in a non-covalent complex. The drug molecule may be a conventional small organic molecule, or a nucleic acid (DNA or siRNA, short interfering RNA, used to silence genes) for non-viral gene delivery, or a conjugated bioactive peptide.

CPPs are transported into the cell by several mechanisms, mainly endocytosis, as discussed below, via receptor-mediated processes, and others. CPPs are able to overcome the Lipinski rule of five (discussed in Section 5.2) in terms of molar mass; indeed, it has been shown that CPPs and PTDs up to 30 residues (i.e. molar mass of a few kDa) can be transported into cells, and also across the blood–brain barrier. The blood–brain barrier is an important barrier that needs to be overcome in order to deliver therapeutics to the brain or nervous system. CPPs are cationic peptides, rich in arginine and/or lysine residues. In this sense they have commonalities with many antimicrobial peptides (AMPs), at least those that act by disrupting (bacterial) cell membranes. CPPs typically have amphipathic α-helical structures, although not always, as in the case of some designed peptides.

Table 4.8 lists several key CPP sequences, along with the origin of the peptide, classed into protein-derived, chimeric (nature-inspired), or model (*de novo*) designed sequences. Designed sequences are generally arginine-rich sequences.

The mechanisms of uptake of CPPs have been extensively examined. The first step of the interaction of CPPs with cells is believed to

Table 4.8   Examples of widely studied cell-penetrating peptides (CPPs).

Peptide	Sequence	Origin
Protein-derived		
Penetratin	RQIKIWFQNRRMKWKK	*Drosophila* antennapedia domain protein
TAT peptide	GRKKRRQRRRPPQ	Human immunodeficiency virus protein
pVec	LLIILRRRIRKQAHAHSK	Derived from murine endothelial cadherin
Chimeric		
Transportan	GWTLNSAGYLLGKINLKALAALAKKIL	Derived from neuropeptide galanin and wasp peptide toxin mastoparan
MPG	GALFLGFLGAAGSTMGAWSQPKKKRKV	Hydrophobic domain from a fusion of HIV gp41 coupled to a SV40 (simian virus 40) T-antigen sequence
Pep-1	KETWWETWWTEWSQPKKKRKV	Derived from an SV40 T-antigen sequence
Designed		
Oligoarginines	$(R)_n$ $n = 6$–12, especially $n = 8$	Model
MAP	KLALKLALKALKAALKLA	Model
R_6W_3	RRWWRRWRR	Model

involve electrostatic interactions with cell-surface polysaccharides such as the glycosaminoglycan (GAG) heparin, whose structure is shown in Figure 4.17, and the closely related heparan sulfate (which has a somewhat different composition of disaccharide units), with both of these being highly anionic. These lead to accumulations of the CPP at the surface of the cell membrane.

CPPs may be translocated into cells through a number of mechanisms, shown in simplified form in Figure 4.18. Direct penetration route may occur via the formation of transient pores or inverted micelle-type structures or may involve adaptive diffusion across the membrane, in which arginine

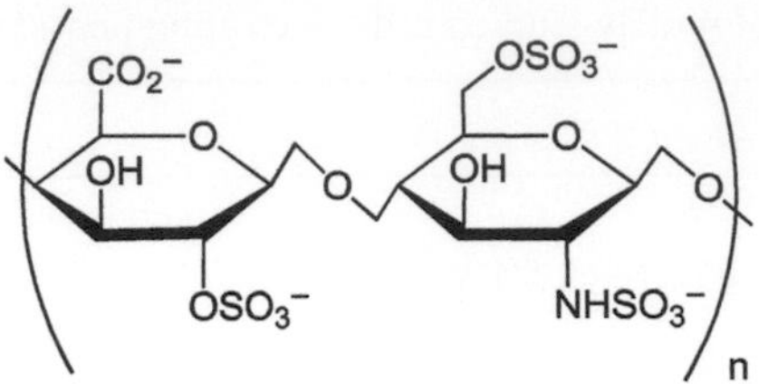

Figure 4.17 Structure of heparin.

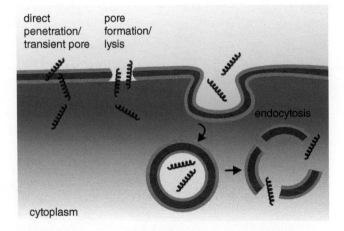

Figure 4.18 Mechanisms of uptake of CPPs.

guanidinium groups form bidentate bonds with lipid phosphate groups (Figure 4.10), masking the peptide charge and enabling transport across the membrane. Alternatively, the CPP may simply cause pore formation or lysis of the cell membrane. The detailed pore formation mechanisms have similar aspects to those shown in Figure 4.16 for AMPs.

Endocytosis is another possible mechanism for CPP uptake. It can occur through caveolin- or clathrin-mediated pathways; these are two types of membrane proteins involved in endocytosis in cells. Macropinocytosis is a distinct mechanism of uptake reported less often for CPPs, in which the peptide is engulfed by the cell membrane without the aid of caveolin or clathrin. By whichever mechanism the peptide penetrates the membrane, once in the cytoplasm, it, along with its cargo, can then diffuse towards its intracellular target, for example the nucleus.

BIBLIOGRAPHY

Bechara, C. and Sagan, S. (2013). Cell-penetrating peptides: 20 years later, where do we stand? *FEBS Letters* 587: 1693–1702.

Brogden, K.A. (2005). Antimicrobial peptides: pore formers or metabolic inhibitors in bacteria? *Nature Reviews Microbiology* 3: 238–250.

Brown, L., Wolf, J.M., Prados-Rosales, R., and Casadevall, A. (2015). Through the wall: extracellular vesicles in Gram-positive bacteria, mycobacteria and fungi. *Nature Reviews Microbiology* 13: 620–630.

Castelletto, V., Barnes, R.H., Karatzas, K.A. et al. (2018). Arginine-containing surfactant-like peptides: interaction with lipid membranes and antimicrobial activity. *Biomacromolecules* 19: 2782–7294.

Cobb, S.L. and Sit, C.S. (2015). Anti-infective peptides. In: *Advances in the Discovery and Development of Peptide Therapeutics* (eds. G. Kruger and F. Albericio). London: Future Science.

Costerton, J.W., Stewart, P.S., and Greenberg, E.P. (1999). Bacterial biofilms: a common cause of persistent infections. *Science* 284: 1318–1322.

Davies, D. (2003). Understanding biofilm resistance to antibacterial agents. *Nature Reviews Drug Discovery* 2: 114–122.

Epand, R.M. and Epand, R.F. (2011). Bacterial membrane lipids in the action of antimicrobial agents. *Journal of Peptide Science* 17: 298–305.

Fux, C.A., Costerton, J.W., Stewart, P.S., and Stoodley, P. (2005). Survival strategies of infectious biofilms. *Trends in Microbiology* 13: 34–40.

Hamley, I.W. (2017). Small bioactive peptides for biomaterials design and therapeutics. *Chemical Reviews* 17: 14015–14041.

Hancock, R.E.W. and Sahl, H.G. (2006). Antimicrobial and host-defense peptides as new anti-infective therapeutic strategies. *Nature Biotechnology* 24: 1551–1557.

Heitz, F., Morris, M.C., and Divita, G. (2009). Twenty years of cell-penetrating peptides: from molecular mechanisms to therapeutics. *British Journal of Pharmacology* 157: 195–206.

Hilpert, K., Fjell, C.D., and Cherkasov, A. (2008). Short linear cationic antimicrobial peptides: screening, optimizing and prediction. In: *Peptide-Based Drug Design* (ed. L. Otvos). Totowa, New Jersey: Humana Press.

Jenssen, H., Hamill, P., and Hancock, R.E.W. (2006). Peptide antimicrobial agents. *Clinical Microbiology Reviews* 19: 491–511.

Kastin, A.J. (2013). *Handbook of Biologically Active Peptides*. San Diego: Academic Press.

Lakemeyer, M., Zhao, W.N., Mandl, F.A. et al. (2018). Thinking outside the box – novel antibacterials to tackle the resistance crisis. *Angewandte Chemie-International Edition* 57: 14440–14475.

Le, C.F., Fang, C.M., and Sekaran, S.D. (2017). Intracellular targeting mechanisms by antimicrobial peptides. *Antimicrobial Agents and Chemotherapy* 61: 16.

Lewis, K. (2001). Riddle of biofilm resistance. *Antimicrobial Agents and Chemotherapy* 45: 999–1007.

Lindgren, M. and Langel, U. (2011). Classes and prediction of cell-penetrating peptides. In: *Cell-Penetrating Peptides: Methods and Protocols* (ed. U. Langel), 3–19. Totowa: Humana Press Inc.

Madigan, M.T., Bender, K.S., Buckley, D.H. et al. (2019). *Brock Biology of Microorganisms*, 15e. New York: Pearson.

Nguyen, J.-T. and Kiso, Y. (2015). Delivery of peptide drugs. In: *Peptide Chemistry and Drug Design* (ed. B.M. Dunn). New York: Wiley.

Phoenix, D.A., Dennison, S.R., and Harris, F. (2013). *Antimicrobial Peptides*. Weinheim, Germany: Wiley-VCH.

Raymond, B. (2019). Five rules for resistance management in the antibiotic apocalypse, a road map for integrated microbial management. *Evolutionary Applications* 12: 1079–1091.

Romling, U. and Balsalobre, C. (2012). Biofilm infections, their resilience to therapy and innovative treatment strategies. *Journal of Internal Medicine* 272: 541–561.

Romling, U., Kjelleberg, S., Normark, S. et al. (2014). Microbial biofilm formation: a need to act. *Journal of Internal Medicine* 276: 98–110.

Stewart, P.S. and Costerton, J.W. (2001). Antibiotic resistance of bacteria in biofilms. *Lancet* 358: 135–138.

Wang, G. (2017). Identification and characterization of antimicrobial peptides with therapeutic potential. In: *MDPI Pharmaceuticals Special Issues*. Basel, Switzerland: MDPI.

Wang, G., Li, X., and Zasloff, M. (2010). A database view of naturally occurring antimicrobial peptides: nomenclature, classification and amino acid sequence analysis. In: *Antimicrobial Peptides: Discovery, Design and Novel Therapeutic Strategies* (ed. G. Wang). Wallingford: CAB International.

Zorko, M. and Langel, U. (2005). Cell-penetrating peptides: mechanism and kinetics of cargo delivery. *Advanced Drug Delivery Reviews* 57: 529–545.

5

Peptide Hormones and Peptide Therapeutics

5.1 INTRODUCTION

The first peptide therapeutic corticotropin was introduced in 1952 and since then more than 70 have been approved for use. Peptide therapeutics are estimated to have sales of more than $15 billion per year, with insulin regulation peptides, growth hormone inhibitors, and GnRH (gonadotropin-releasing hormone) modulators being particularly important classes. Most approved peptides are used to treat cancer, metabolic diseases, fertility disorders, and fertility regulation, and are used in haematology, urology, and bone disease (Figure 5.1). The pipeline of peptides in development has a very different emphasis, with particularly large numbers of trials in progress for peptides with anticancer or antimicrobial activity as well as metabolic and cardiovascular diseases.

The development of peptide therapeutics is an exciting field, with considerable investment and research from the pharmaceutical industry which has led to, for example, the Federal Drug Administration (FDA) approving 17 peptide therapeutics for the market in the period 2000–2010. Hundreds of peptides are the subject of ongoing clinical trials.

The chemistry and modes of action of many of these peptides are discussed in the following. A variety of modifications of the peptides to ensure longer stability *in vivo* are available, including examples of lipidated peptides, those with substitutions that reduce cleavage of specific residues, D-amino acid-based peptides, and conjugates to albumin or antibody fragments.

Introduction to Peptide Science, First Edition. Ian W. Hamley.
© 2020 John Wiley & Sons Ltd. Published 2020 by John Wiley & Sons Ltd.

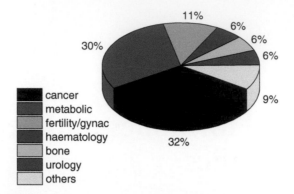

Figure 5.1 Number of approved peptide therapeutics according to treatment field. Some peptides have indications across more than one of these classifications.

Around one third of recently introduced peptide therapeutics, including those under development, are conjugates of peptides.

Peptide hormones are vital agents in many biological processes, and many peptide therapeutics are derived from peptide hormones. This chapter discusses the main classes of peptide hormones, along with a summary of the main peptide therapeutics.

This chapter is organized as follows. Section 5.2 presents some general considerations concerning the use of peptides as therapeutic molecules. Section 5.3 describes peptide hormones in detail, with subsections devoted to the main classes. Neuropeptides are discussed in Section 5.4, while venom-derived peptides are the subject of Section 5.5, and anticancer peptides are introduced in Section 5.6. Section 5.7 covers miscellaneous bioactive peptide agents. Section 5.8 concludes the chapter with a brief outline of peptides and lipopeptides used in cosmetic (skincare) formulations.

5.2   GENERAL PRINCIPLES OF PEPTIDE THERAPEUTICS

The largest class of peptide therapeutics are based on peptide hormones. These are discussed in the following sections according to the type of hormone, classified according to the gland or organ where it is secreted. A particular target for peptide therapeutics is G protein-coupled receptors (GPCRs). Indeed, 40% of peptides that have entered the clinic since 2010 have targeted GPCRs. Other cell-surface receptors such as natriuretic peptide receptors and cytokine receptors are common targets.

It may seem surprising in some respects that peptides, as relatively high molar mass compounds, are potentially valuable as therapeutics.

Lipinski's rule of five for 'drugability' used for low molecular weight compounds suggests an upper limit of $500\,\mathrm{g\,mol^{-1}}$ for such therapeutics as well as limitations on the number of hydrogen bond donors and acceptors and octanol/water partition coefficient (log$P$). For peptides the 'upper limit' of $500\,\mathrm{g\,mol^{-1}}$ is circumvented *in vivo* because peptides are natural compounds which are essential components of many biological processes as hormones and other agents. As such, there are numerous pathways by which peptides can be taken up into cells via specific receptors and transduction mechanisms. They can act extracellularly through interactions with specific proteins, for example antibodies (immunoglobulins) in the immune system. There are of course many binding mechanisms by which proteins interact with peptides. Despite the compatibility of peptides with various receptors, *in vivo* stability is an issue due to proteolysis (degradation by enzymes such as proteases and peptidases). Strategies to improve the stability and extend the *in vivo* circulation (half-life) of peptides, including cyclization, lipidation, PEGylation, and others, are discussed in Sections 2.6–2.14.

Peptides offer important advantages as well as drawbacks. Some key factors are summarized in a SWOT analysis shown in Figure 5.2.

One advantage of peptide therapeutics is that it is possible to screen sequences via automated synthesis methods, for example alanine scans (substitution of an alanine at each residue position) may be performed or specific

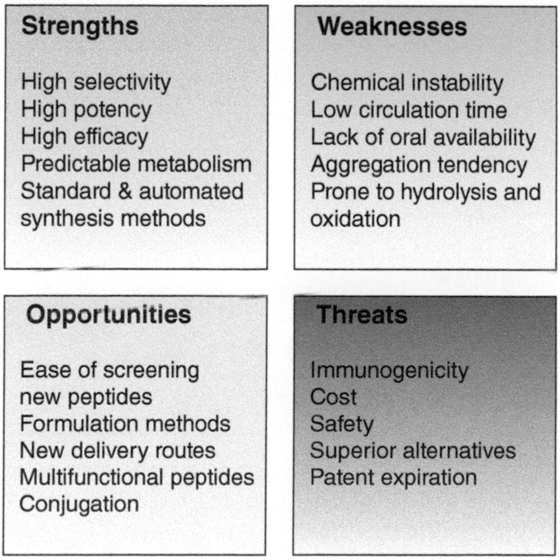

Strengths	Weaknesses
High selectivity High potency High efficacy Predictable metabolism Standard & automated synthesis methods	Chemical instability Low circulation time Lack of oral availability Aggregation tendency Prone to hydrolysis and oxidation
Opportunities	**Threats**
Ease of screening new peptides Formulation methods New delivery routes Multifunctional peptides Conjugation	Immunogenicity Cost Safety Superior alternatives Patent expiration

Figure 5.2 SWOT analysis of peptide therapeutics.

substitutions explored. Sequence randomization (for example split-mix synthesis discussed in Section 2.5) can also be useful when exploring peptide space in terms of structure–activity relationships (SARs).

Due to their tendency to degrade *in vivo*, oral delivery of peptides is challenging. The pharmacokinetic (PK) profile of any drug is very important, but degradation of peptides can be problematic. Pharmacokinetics relates to the adsorption, distribution, and metabolism of drugs. It should be distinguished from pharmacodynamics which refers to the study of the effect of the drug on the organism. Peptide therapeutics are most commonly delivered via injection or subcutaneous administration, although nasal sprays are available for some, including formulations of calcitonin, desmopressin, and oxytocin (with other examples discussed in Section 5.3.1) and a number can be delivered orally. Figure 5.3 shows delivery routes for marketed peptides. The most successful parenteral formulations depend on slow release systems, mainly using PLGA [poly(lactic-co-glycolic acid)] polymer matrices. In the future, needle-free transdermal and microneedle technologies are likely to become increasingly widespread. These use gas- or air-jets instead of a hypodermic needle to penetrate the skin. Topical delivery can be enhanced using ultrasound or iontophoresis, which is the use of a low-level current to assist delivery.

'Oral bioavailability' refers to the quantity of a therapeutic that is absorbed after oral administration. The oral bioavailability of many peptide drugs is typically only 1–2% due to the barriers within the gastrointestinal (GI) tract. Molecules can move across the intestinal epithelium either through cells (transcellular transport) or between cells (paracellular transport). Transcellular transport mechanisms include endocytosis, carrier-mediated transport, or passive diffusion. The transport of peptide drugs via one of these mechanism is hindered by their high molecular weight and their usually low lipophilicity. Peptide drugs are also prone to

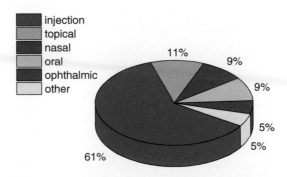

Figure 5.3   Delivery routes for marketed peptides.

enzymatic degradation by proteases, aminopeptidases, endopeptidases, dipeptylpeptidases, etc. The mucus layer of the intestine wall also provides a physical barrier against diffusion. Finally, there are pH variations within the GI tract which influence peptide stability.

For peptide therapeutics, oral bioavailability can be enhanced via strategies described in Sections 2.8–2.11, including lipidation, glycosylation, and PEGylation as well as formulation with excipients to improve permeability. These chemical modifications can all reduce the degradation of peptides *in vivo* (by proteases) and lipidation or glycosylation can improve permeability across membranes and/or enable targeting of specific receptors. As an alternative, adsorption/permeation-enhancing agents, including bile salts, fatty acids, surfactants, chelating agents, efflux inhibitors, and polymers, may be used to enhance transport across the GI tract. Particulate carrier systems (micro- and nano-particles) may be employed; examples include liposomes and polymer capsules. Mucolytic agents such as N-acetyl-L-cysteine (Section 5.7) may also be used. As an alternative strategy for oral delivery, peptides may be targeted towards peptide transporters such as PepT1 located in the intestinal epithelium, this being able to transport di- and tri-peptides and peptidomimetic analogues. Enzyme inhibitors may be employed; some of these, such as amastatin, boroleucine, leupeptin, pepstatin (all shown in Figure 5.4), and aprotinin (a 58-residue peptide) are peptide derivatives. Mucoadhesive materials such as chitosan or poly(acrylic acid) can be used to extend the adsorption residence time at the epithelium.

The blood–brain barrier (BBB) represents another major challenge in the development of peptide therapeutics. The BBB is an endothelial barrier in brain microvascular capillaries that protects the brain from unwanted circulating pathogens, but it needs to be circumvented by a successful therapeutic in the treatment of CNS (central nervous system) diseases. Transport across the BBB occurs via paracellular or transcellular mechanisms, as in the GI tract discussed above. The tight junctions between epithelial cells are disrupted in neurodegenerative disorders, including Alzheimer's disease and bacterial meningitis, enabling transcellular transport. Polar solutes, including amino acids and small peptides, are able to cross the BBB via solute carrier-mediated transport. Active efflux carriers, including ABC transporters (ATP-binding cassette transporters), can also pump substances from endothelial cells to blood or brain. Amphiphilic and lipid-based drugs are transported through this route. Lipidation enhances the ability of a peptide drug to diffuse across the BBB, as does halogenation. On the other hand, high polarity or high hydrogen bonding capacity are associated with reduced BBB penetration via diffusion. The BBB can be bypassed via the L-system which is used to increase levodopa levels. Some amino acids and derivatives

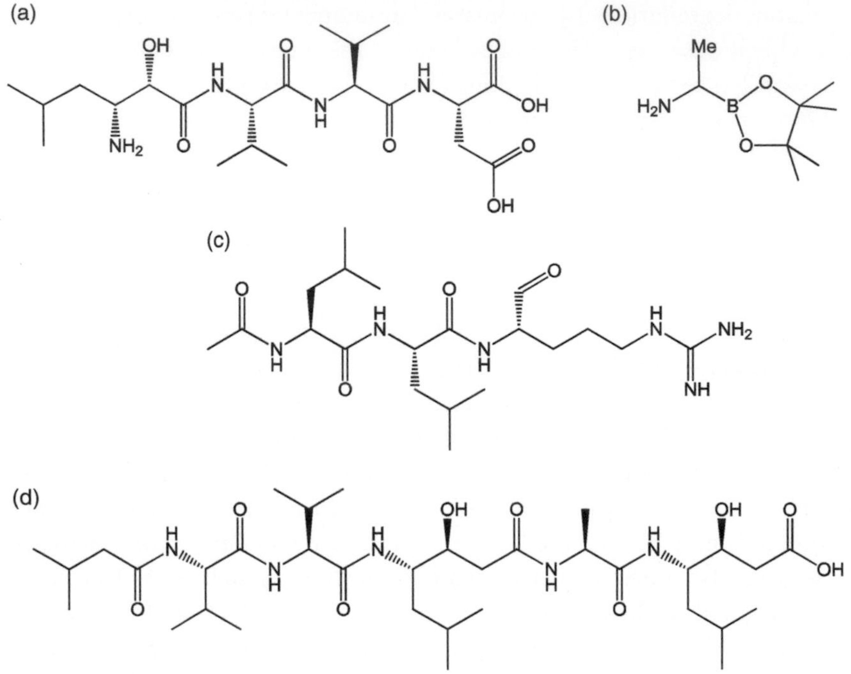

Figure 5.4 Examples of peptide enzyme inhibitors: (a) amastatin, (b) boroleucine, (c) leupeptin, (d) pepstatin.

can be transported by the L-system including α-methyldopa, GABA (see Table 1.3), and analogues. Glycosylation enhances BBB permeability due to increased endocytotic transcellular transport. Formulation of peptides in liposomes has also been used to deliver them across the BBB. Another means to transport peptides across the BBB is to prepare analogues of natural BBB-permeating peptides such as the opioid enkephalin peptides (Section 5.4). The approach of conjugating peptides to moieties that target receptors in the brain has been used when attaching transferrin antibodies conjugated to peptides. The Tat cell-penetrating peptide (see Section 4.11) is also able to cross the BBB and can carry heterologous peptides with it. Instead of modifying the peptide drug, the permeability of the BBB itself can be increased by modulating the expression of transporters and/or their signalling methods or by osmotic or ultrasound-induced opening of the BBB. Note, however, that it is risky to disrupt this protective barrier of the brain. Drugs can also be delivered across the BBB via neurosurgical methods. Intranasal delivery is a potentially more attractive method to deliver a drug, avoiding the BBB, and has been shown to increase levels of peptide in cerebrospinal fluid.

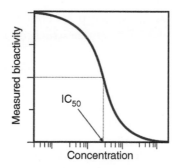

Figure 5.5 Schematic of a dose–response curve, showing bioactivity versus concentration plotted on a logarithmic scale, with IC_{50} value indicated.

The potency of a peptide as a therapeutic for a particular indication is commonly quantified in terms of IC_{50} values. This is the half maximal inhibitory concentration, i.e. the concentration (dose) of peptide required to reduce the measured biological activity by 50%. This can be obtained from a dose–response curve (Figure 5.5), which is the most common pharmacodynamic measurement.

Peptides may be developed as therapeutics using quantitative structure activity relationship (QSAR) modelling, which is employed for small molecule drugs. This method develops a set of models to relate physicochemical or theoretical properties of the drug to the response (in this case the particular bioactivity of the peptide), and to build on this to provide predictive capability.

Peptides can also be used as leads for drug discovery. After screening of peptide candidates for activity, derivatives can be prepared with greater *in vivo* stability or greater bioavailability. The palette of modifications available include substitutions with D-amino acids or non-natural residues and/or protecting groups and strategies such as those used to produce peptidomimetics, as described in Section 2.7. These include N-methylation or formation of other peptoids, or incorporation of β- and γ-amino acids. Cyclization can also be used to increase stability against proteolytic enzymes. Cyclization is discussed further in Section 2.6. Incorporation of non-natural residues and/or D-amino acids reduces degradation by proteases (endopeptidases). Non-natural residues can include halogenated aromatic residues, which reduce the recognition of the peptide by the enzymatic cleavage site. Depsipeptides also show lowered degradation by proteolysis. Glycosylation or lipidation can increase stability by reducing degradation and/or facilitating binding to serum proteins such as albumin. A further set of strategies to prepare peptide-based drugs employs the use of isosteres other than peptidomimetic structures. Isosteres are structures with the same number of atoms/valence electrons as the starting peptide, but arranged differently or with different bond types. Thioamide structures are one example among many other classes of pseudo-peptide.

An important issue to consider when developing a peptide for therapeutic activity is the need to maintain exceptionally high purity of the product and to ensure its stability in storage and in use. Degradation can occur by denaturation, proteolysis, oxidation and reduction, disulfide exchange, microbial activity, and racemization. Aggregation and precipitation may also be problematic. Solubility also needs to be considered. Effects of interactions with excipients and even the drug container need to be investigated. All these issues are carefully examined as part of the regulatory approval process, for example FDA approval in the USA.

5.3   PEPTIDE HORMONES

A very large number of hormones are produced in the body, including peptide hormones, protein hormones, and steroid hormones. Peptide hormones are secreted naturally and may be used as therapeutics directly (synthetic or recombinant versions of natural peptides) or as the basis to design related peptides (or peptidomimetics) such as those that target specific biological processes. Peptides are stored in secretory granules, often in the form of amyloid structures. For example, pituitary gland secretory granules have been observed . Peptide hormones and neuropeptides pack into dense-core vesicles (DCVs), which are amyloid-like aggregates used to temporarily store peptide messengers in secretory cells. When DCVs are triggered, they release the stored contents into the blood or extracellular space. Therefore for these types of peptides, reversibility of peptide aggregation is essential for their function.

Peptide hormones primarily have an effect on the endocrine system. The endocrine system is composed of many different glands and it can be divided into two categories; classical and non-classical. In the endocrine system, hormones are secreted into the circulatory system where they are distributed throughout the body, regulating bodily functions. The major endocrine glands include the pituitary gland, pineal gland, pancreas, thyroid gland, adrenal glands, and ovaries/testes. The primary function of these glands is to manufacture specific hormones. Other endocrine-secreting organs include the thymus, heart, hypothalamus, kidneys, liver, and the gastrointestinal tract. Many of the classical hormones are controlled by the hypothalamus and pituitary; these organs can also be classified as being an extension of the nervous system.

Peptide hormones act on the surface of target cells at receptors such as GPCRs via second messengers (Figure 5.6) in the cytoplasm that stimulate

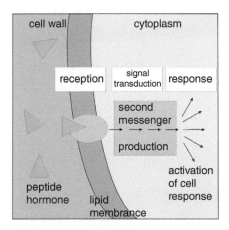

Figure 5.6   Action of peptide hormones at cell-surface receptors.

cellular responses such as enzyme activation or gene transcription, for example via the ATP/cAMP-dependent pathway (ATP is adenosine triphosphate, an essential biological energy source, cAMP is cyclic adenosine monophosphate, a second messenger nucleotide). This mode of action differs from steroid hormones, as these are lipid soluble, and so can move through the plasma membranes of target cells and act within the nuclei.

Many peptide hormones are available as native therapeutic agents or are used to produce synthetic derivatives. Table 5.1 shows some of the leading peptide hormone therapeutics. Peptide hormones are often delivered intravenously, due to stability issues. Conjugation of peptides to lipids, glycopolymers, or PEG (polyethylene glycol) increases circulation time. However, conjugation does not fully overcome the intrinsic propensity of peptides to be degraded by proteases, which occurs in the digestive tract if taken orally. As one example, the highest selling marketed diabetic drug Liraglutide incorporates a lipid chain to extend plasma circulation; however, it has to be taken via subcutaneous injection.

5.3.1   Hypothalamus and Anterior Pituitary (Adenohypophysis) Hormones, Including Growth Hormones and Growth Hormone Inhibitors

The hypothalamus and the closely related anterior pituitary gland in the brain control much of the endocrine system. The neurons in the hypothalamus synthesize important release hormones (or factors) and release-inhibiting hormones discussed in this section, which upon delivery to the anterior pituitary release trophic hormones (those stimulating activity of a

Table 5.1 Examples of established peptide hormone-based therapeutics on the market.

Peptide	Trade name	Activity	Molecular properties
Insulin	(Native peptide)	Treatment of types I and II diabetes	Peptide/small protein containing two chains (21-residue A chain and 30-residue B chain)
Glucagon	(Native peptide)	Hypoglycaemia in diabetic patients	29-residue peptide cleaved from proglucagon
Adrenocorticotropic hormone (ACTH)	(Native peptide)	Activity against adrenal diseases, e.g. Cushing's disease, Addison's disease. Used to treat epileptic spasms in infants, multiple sclerosis in adults, psoriatic arthritis, rheumatoid arthritis, ankylosing spondylitis, and others including skin conditions	39-residue peptide obtained by post-translational modification and cleavages of pro-opiomelanocortin
Tetracosactide	N/A	ACTH stimulation test	24-residue N-terminal sequence of ACTH
Cyclosporin, ciclosporin	N/A	Immunosuppressant, rheumatoid arthritis, psoriasis, Crohn's disease, nephrotic syndrome, prevention of rejection in organ transplants, conjunctivitis	Natural fungal 11-residue cyclic peptide/peptoid hybrid
Leuprorelin, leuprolide	Lupron	Prostate cancer, breast cancer, endometriosis, uterine fibroids, early puberty	Synthetic nonapeptide analogue of naturally occurring gonadotropin-releasing hormone (GnRH)
Growth hormone-releasing hormone (GHRH)	Somatorelin	Diagnostic for growth hormone deficiency	Native 44-residue peptide
Tesamorelin	Egrifta	HIV-associated lipodystrophy	GHRH analogue with N-terminal *trans*-hexenoic acid

Cetrorelix	Cetrotide	In vitro fertilization to modulate hormone release	Synthetic decapeptide analogue of naturally occurring gonadotropin-releasing hormone (GnRH)
Goserelin	Zoladex	Breast and prostate cancer	GnRH analogue decapeptide with two substitutions to inhibit rapid degradation
Degarelix	Firmagon	Prostate cancer	GnRH analogue decapeptide (GnRH antagonist)
Calcitonin	(Native peptide)	Osteoporosis and other bone diseases, hypercalcaemia	32-amino acid polypeptide, salmon calcitonin used therapeutically
Liraglutide	Vicoza	Type 2 diabetes, obesity	97% homologous to native human GLP-1 (7–37) by substituting arginine for lysine at position 34 and addition of a fatty acid chain
Octreotide	Sandostatin	Acromegaly (excess growth hormone disorder), hypoglycaemia, gastrointestinal fistulae	More potent synthetic octapeptide analogue of naturally occurring somatostatin
Teriparatide	Forteo	Osteoporosis	Recombinant form of parathyroid hormone consisting of the bioactive N-terminal 34 amino acids
Exenatide	Byetta	Type II diabetes	GLP-1 agonist. Synthetic version of exendin-4, a hormone found in the saliva of the Gila monster reptile
Lanreotide	Somatuline	Acromegaly	Cyclic peptide that is a long-acting analogue of somatostatin
Vasopressin	(Native peptide)	Antidiuretic hormone deficiency, heart conditions	Partly cyclic nonapeptide, native peptide
Oxytocin	(Native peptide)	Induction of labour, control of postpartum bleeding, promotion of lactation	Natural cyclic peptide hormone

(continued)

Table 5.1 (*continued*)

Peptide	Trade name	Activity	Molecular properties
Demoxytocin	Sandopart, Odeax, Sandopral	Induction of labour, promotion of lactation, prevention of postpartum mastitis	Analogue of oxytocin in which Cys is replaced with Mpa = β-mercaptopropionic acid
Atosiban	Tractocile, Antocin, Atosiban SUN	Prevention of premature labour	Nonapeptide, desamino acid analogue of oxytocin. Vasopressin/oxytocin receptor agonist.
Carbetocin	Duratocin, Pabal, Lonactene, Depotocin, Comoton, Decomoton	Control of postpartum bleeding	Octapeptide analogue of oxytocin
Angiotensin II		Control of blood pressure after septic shock	Natural linear octapeptide
Thymosin α₁ (thymalfasin)	Gipreza Zadaxin	Hepatitis B and C and immune-system boosting adjuvant	28-residue linear peptide

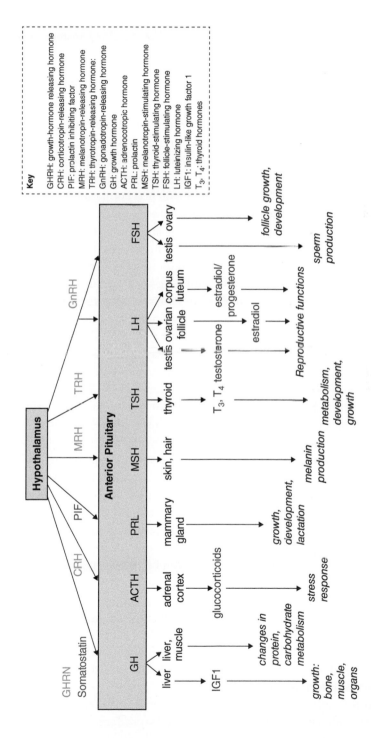

Figure 5.7 The hypothalamic–pituitary system, showing key hormones and their target organs and activities. Hypothalamic releasing hormones are indicated in blue and their release-inhibiting factors in red. Effector molecules are shown in purple.

gland) which then stimulate specific endocrine tissues. Figure 5.7 shows a schematic overview of the hypothalamus/anterior pituitary system, along with key hormones and their roles. The hormones which are discussed in the following sections are peptides, with the exception of TSH, FSH, and LH, which are glycoproteins, and PRL, which is a (198-residue) protein. Abbreviations are defined in Figure 5.7.

Hypothalamic releasing hormones are formally known as '-liberins', for example corticoliberin, and the corresponding inhibiting peptides are known as '-statins', such as somatostatin. These are IUPAC recommendations, although these terms are not widely encountered. The latter '-statins' are not to be confused with lipid-lowering medicinal compounds.

Adrenocorticotropic hormone (ACTH) is a native peptide hormone, introduced in the 1950s as a therapeutic agent to treat various conditions characterized by over- or under-production of ACTH or deficiency of cortisol. It is a 39-residue peptide, the first 13 of which (counting from the N-terminus) may be cleaved to form α-melanocyte-stimulating hormone (α-MSH). ACTH is produced by the anterior pituitary gland in response to stress, stimulating the production of glucocorticoid steroid hormones such as cortisol. ACTH is a lipotropin produced by the cleavage of pro-opiomelanocortin (POMC). The other hormone produced in this way is the 90-residue peptide β-lipotropin that can further be cleaved to γ-lipotropin that, in turn, can be cleaved to give α-MSH. ACTH from pig pituitary glands is a therapeutic known as corticotropin and is used for a wide range of conditions resulting from dysfunction of the adrenal cortex, including multiple sclerosis (MS), collagen diseases, inflammatory rheumatoid diseases, and others (see also Table 5.1). Tetracosactide is the 24-residue N-terminal sequence of ACTH and this peptide, also known as tetracosactrin or cosyntropin, is used in the ACTH stimulation test to examine the function of the adrenal glands.

There are three forms of melanocyte-stimulating hormones (MSH hormones or melanotropins) from the melanocortin family, α-, β-, and γ-MSH. Human α-MSH is a 13-residue peptide with sequence Ac-SYSMEHFRWGKPV-NH₂ while human β-MSH has sequence AEKKDEG-PYRMEHFRWGSPPKD, and γ-MSH is YVMGHFRWDRFG. Peptide α-MSH has a role in melanin production, appetite suppression, and sexual arousal. Synthetic analogues of α-MSH have been developed as photoprotective agents, in particular afamelanotide, Ac-SYS(Nle)EHfRWGKPV-NH₂ (Nle is norleucine, 2-aminohexanoic acid, see Table 1.3). Afamelanotide is used to prevent skin damage from the sun in patients with erythropoietic protoporphyria. It is injected as an implant.

Corticotropin-releasing hormone (CRH or cortico-releasing factor) is a 41-residue hypothalamic peptide that stimulates production of ACTH as part of the stress response. Its deficiency can cause hypoglycaemia. It is also highly expressed in the placenta as a marker that senses length of gestation and the timing of parturition and delivery. Corticorelin is a recombinant version of CRH used for diagnostics.

GnRH (gonadotropin-releasing hormone, also known as gonadorelin or luteinizing hormone-releasing hormone, LHRH) is produced in the hypothalamus from where it is transported to the pituitary gland, where it controls the production of luteinizing hormone (LH, a heterodimeric glycoprotein) and follicle-stimulating hormone (FSH). This process affects the release of steroid sex hormones such as testosterone and progesterone. GnRH forms the basis of an important family of therapeutics, the GnRH modulators, agonists, and antagonists. These are also known as LHRH modulators or LHRH receptor modulators. As shown in Figure 5.8a, GnRH is a decapeptide with sequence (pG)HWSYGLRPG-NH$_2$ (pG is pyroglutamic acid, see Table 1.3). This peptide, produced in the hypothalamus, binds to the LHRH receptor and triggers the release of LH and FSH, which are then responsible for the production of testosterone and sperm, respectively, in men. In females, LH and FSH stimulate ovarian follicular growth prior to ovulation and changes in their levels during menstruation. These hormones are also involved in the development of male and female sexual characteristics during puberty. Another trophic peptide hormone produced in the hypothalamus is TRH (thyrotropin-releasing hormone, Figure 5.8b). This tripeptide (L-proline-L-histidyl-L-pyroglutamyl amide) contains the pyroglutamyl

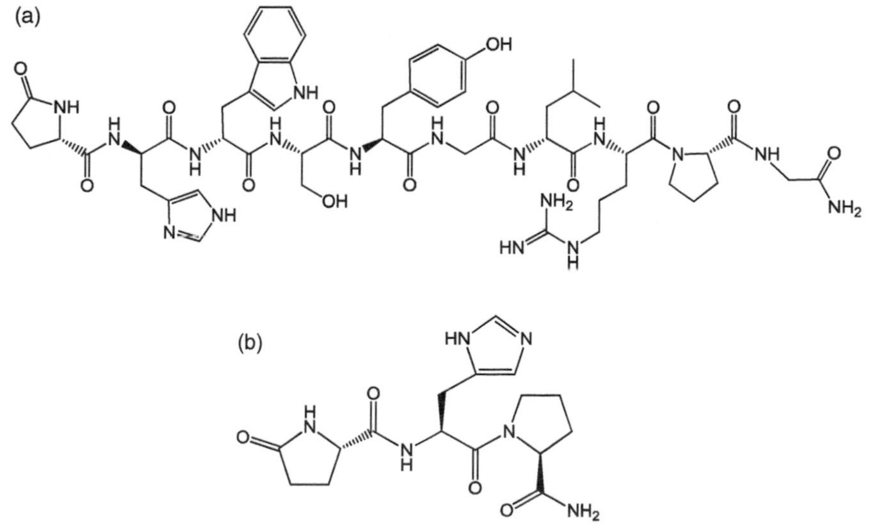

Figure 5.8 Molecular structure of (a) GnRH and (b) thyrotropin-releasing hormone.

residue. It has the commercial name protirelin and is used in the diagnosis of thyroid disorders and to treat spinocerebellar ataxia, among other applications. Taltirelin is a TRH analogue which can also be used to treat spinocerebellar ataxia.

GnRH agonists include buserelin, gonadorelin, goserelin, histrelin, leuprorelin (leuprolide), nafarelin, and triptorelin, which are delivered by injection (except buserelin and nafarelin taken in nasal sprays). Currently approved GnRH antagonists include the peptide molecules cetrorelix, abarelix, degarelix, and ganirelix. The sequences of GnRH agonist peptides are summarized in Table 5.2. The GnRH antagonist sequences are exemplified by that of cetrorelix shown in Figure 5.9.

Cetrorelix is a decapeptide GnRH analogue which was used to treat breast, ovarian and prostate cancer, endometriosis and uterine fibroids, however use for these applications has been discontinued and it is now used in the process of *in vitro* fertilization. It is a GnRH antagonist. Cetrorelix has the

Table 5.2 Sequences of GnRH agonists with generic sequence (pG)HWSYxLRPy.

Peptide	Sequence[a]		Indications
	x	y	
GnRH	Gly	Gly-NH$_2$	Natural hormone
Buserelin	D-Ser (OtBu)	-NHEt	Prostate cancer, breast cancer, endometriosis, infertility, uterine fibroids
Gonadorelin	Gly	Gly-NH$_2$	Identical to GnRH sequence. Amenorrhea, delayed puberty, infertility
Goserelin	D-Ser(OtBu)	AzGly-NH$_2$	Prostate cancer, breast cancer
Histrelin	D-His(N-benzyl)	-NHEt	Prostate cancer, uterine fibroids, precocious puberty
Leuprorelin	D-leu	-NHEt	Prostate cancer, breast cancer, endometriosis, uterine fibroids, precocious puberty
Nafarelin	2-Nal	Gly-NH$_2$	Endometriosis, uterine fibroids, precocious puberty, *in vitro* fertilization, testosterone suppression in transgender women
Triptorelin	D-Trp	Gly-NH$_2$	Prostate cancer, endometriosis, uterine fibroids, male hypersexuality

[a]OtBu denotes *tert*-butyloxycarbonyl, AzGly denotes aza-glycine, and 2-Nal denotes 2-napthylalanine (see Table 1.3).

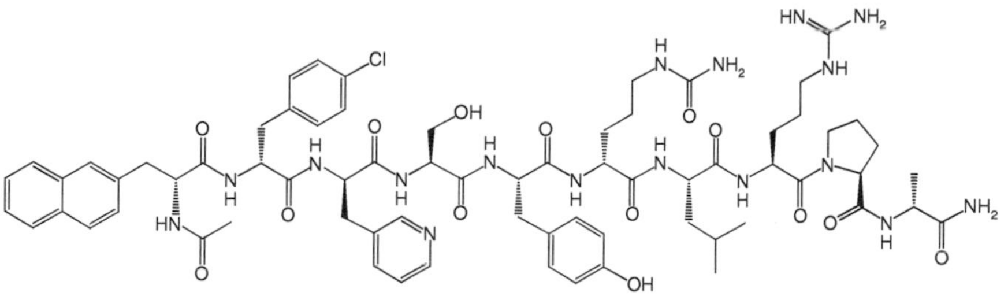

Figure 5.9 Molecular structure of cetrorelix.

sequence Ac-D-2Nal-D-Cpa-D-3Pal-Ser-Tyr-D-Cit-Leu-Arg-Pro-D-Ala-NH$_2$, containing a number of non-canonical residues including 2-Nal = 2-naphthylalanine, Cpa = chlorophenylalanine, 3Pal = [3-(3-pyridylalanine)], and Cit = citrulline (Table 1.3). Abarelix, degarelix, and ganirelix have related sequences with non-natural residues at positions 5 and 6 (and 8 in the case of ganirelix in which Arg is replaced by homoArg, Table 1.3). Abarelix and degarelix, like cetrorelix, reduce testosterone production (via LH and FSH) and are treatments for prostate cancer. Ganirelix is used in fertility treatment and for endometriosis.

GHRH (growth hormone-releasing hormone) is a larger 44-residue peptide hormone, not to be confused with GnRH. It is produced in the hypothalamus and is also known as somatocrinin or somatorelin in its medical form. Tesamorelin is a variant comprising the native 44-residue sequence with an added N-terminal hexenoic acid chain. GHRH is also used as a diagnostic for growth hormone deficiency.

Somatostatin, also known as growth hormone-inhibiting hormone (GHIH), is a 14-residue cyclic peptide (Figure 5.10a). It plays an important role in cell proliferation (via inhibition of the secretion of growth factors and cytokines) and neurotransmission.

Lanreotide and octreotide (Figure 5.10b) are synthetic analogues of somatostatin with modifications shown schematically in Figure 5.10c. These cyclic peptides are used to treat acromegaly (a condition where the body produces too much growth hormone). They both are octapeptides and have greatly extended circulation times compared to native somatostatin. Both octreotide and lanreotide act on the same receptors as somatostatin (G protein-coupled somatostatin receptors) and both are inhibitors of insulin and glucagon. These peptides were designed to have enhanced *in vivo* stability due to the inclusion of D-amino acids as well as other modifications. The D-configuration of tryptophan fosters a bioactive conformation, based on a β-turn centred on the Trp-Lys motif (Figure 5.10).

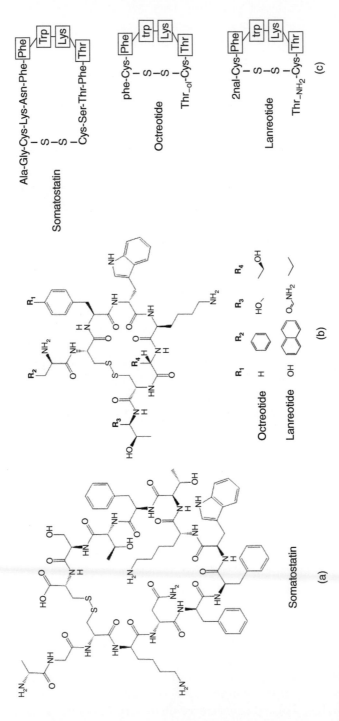

Figure 5.10 Molecular structures of: (a) somatostatin, (b) octreotide and lanreotide, (c) cartoon representation showing common features. phe and trp denote D-amino acids, 2nal refers to D-2-napthylalanine, Thr$_{-ol}$ is threoninol and Thr$_{-NH_2}$ is threoninamide.

In water, lanreotide self-assembles into monodisperse nanotubes (see Section 3.10.1) that form liquid crystalline phases. The nanotubes are made up of dimers that self-assemble into a 2D crystal that is held together by lateral chain interactions and also by antiparallel β-sheet formation. This self-assembly behaviour into liquid crystalline structures is the basis of the Somatuline Autogel slow-release system to treat acromegaly.

Pasireotide is a cyclic hexapeptide containing several non-natural residues that was also designed as a somatostatin analogue. It is an orphan drug (one developed to treat a rare condition) used for Cushing's disease, which is a cause of Cushing's syndrome that is characterized by increased secretion of ACTH.

5.3.2   Thyroid Hormones

Calcitonin is a 32-residue peptide (containing one disulfide bridge) produced by parafollicular cells in the thyroid. It has a key role in the reduction of calcium levels (calcium homeostasis) in serum. Calcitonin is an example of a short peptide with a α-helical structure. Calcitonin is used therapeutically (in the form of salmon calcitonin) to treat osteoporosis and hypercalcaemia. The action of calcitonin is opposed by parathyroid hormone (PTH, parathormone or parathyrin), a 90-residue peptide pro-hormone. PTH regulates serum calcium and plays a role in many bone diseases, in hyperparathyroidism and hypoparathyroidism, in hypercalcaemia and hypocalcaemia (referring to calcium levels in serum), and in regulation of vitamin D production in the kidney.

Elcatonin is a calcitonin derivative in which the disulfide bridge is replaced with a more stable alkyl chain. It is used to treat osteoporosis. Teriparatide is a 34-residue peptide formed from the N-terminal fragment of PTH. Recombinant teriparatide is sold under the trade name Forteo. It is used to treat osteoporosis. Abaloparatide, FDA approved in 2017, is a second-generation version of teriparatide, with multiple amino acid substitutions. It also activates the PTH receptor PTH1R and is a bone growth promotor, used in the treatment of osteoporosis.

Etelcalcetide is a peptide treatment for secondary hyperparathyroidism for patients undergoing haemodialysis. It targets the calcium-sensing receptor (CaSR) in the parathyroid gland. The peptide is a D-amino acid peptide (rich in D-Arg) with a disulfide bridge between the D-cysteine residue and L-cysteine amino acid (Figure 5.11).

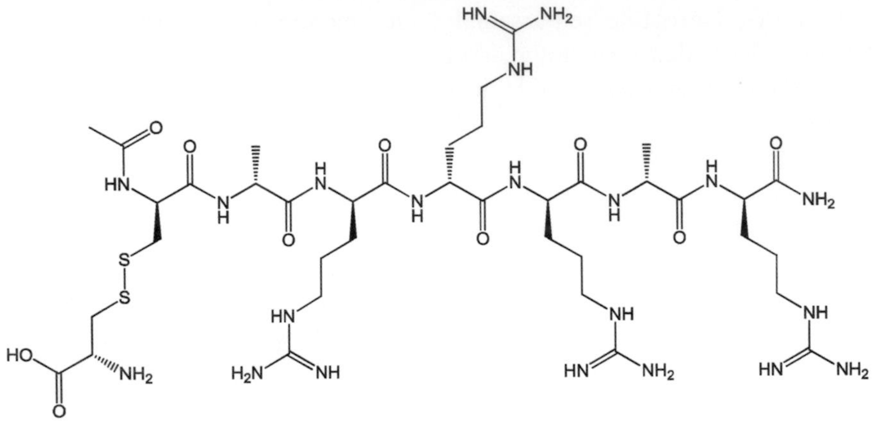

Figure 5.11   Molecular structure of etelcalcetide.

Thyroxine (T_4) and triiodothyronine (T_3) are two tyrosine derivatives (i.e. they are amino acid derivatives) involved in metabolism. Their production is stimulated by TSH (see Figure 5.7).

5.3.3   Pancreatic Hormones

The peptide hormones insulin and glucagon, which are both produced in the pancreas, play an interdependent role in the regulation of blood sugar via pathways illustrated schematically in Figure 5.12. High blood sugar levels (as a result, for example, of eating a carbohydrate-rich meal) stimulate the

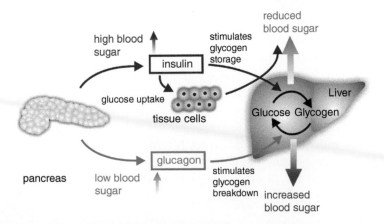

Figure 5.12   Pathways by which insulin and glucagon regulate blood sugar levels.

pancreatic beta cells to release insulin. Some of the blood sugar is taken up by body cells (leading to lowered blood sugar) and some of is converted into glycogen and stored in the liver. Low blood sugar levels (e.g. after fasting), on the other hand, stimulate release of glucagon from the alpha cells of the pancreas. The glucagon stimulates glycogen breakdown and release of glucose into the blood. Together, these processes should stabilize glucose to produce homeostatic normal blood glucose levels of around 70–140 mg/100 ml (lower range before eating, upper range after). Diabetics, of course, suffer from insufficient insulin production which disturbs the pathways shown in Figure 5.12 and leads to altered blood sugar levels.

Insulin is taken by injection to treat type 1 and type 2 diabetes mellitus. Insulin is a two-chain peptide hormone comprising 21-amino acid A chain bound (via two disulfide bridges) to a 30-residue B-chain (Figure 5.13). It is secreted by beta cells in the pancreatic islets (regions of the pancreas containing hormone-releasing cells). It was the first therapeutic peptide hormone, being introduced to the clinic in the 1920s.

A number of derivatives of insulin with extended circulation time are currently marketed as injectables. These are shown in Figure 5.14. Insulin glargine has a substitution in the A-chain at N21 and a C-terminal extension of two arginine residues in the B-chain. This shifts the isoelectric point, leading to greater solubility at acidic pH, but lower solubility at physiological pH. This enables the formulation of a clear injectable solution. Insulin glargine is among the top 25 most prescribed drugs in the US. Other insulin derivatives involve substitutions and extensions in the B-chain. The C-terminal P and K residues are switched in insulin lispro, which is produced recombinantly. This does not modify the receptor binding, but prevents the

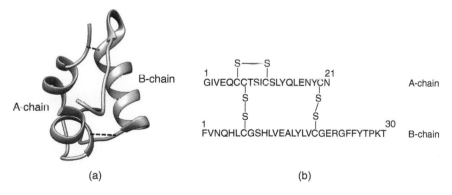

(a)   (b)

Figure 5.13   (a) Three-dimensional crystal structure of insulin, showing inter-chain disulfide bonds (there is also a disulfide loop within the A chain, shown right). (b) Chemical structure.

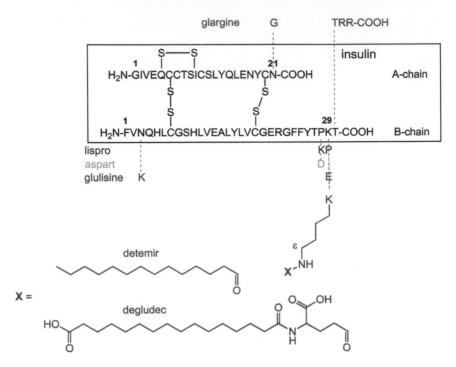

Figure 5.14　Insulin analogue peptides on the market.

formation of insulin multimers (especially dimers and hexamers) so that more monomeric insulin is available after injection. Insulin aspart involves substitution of the C-terminal P for D; this lowers the isoelectric point and leads to a faster acting form. Insulin glulisine is another fast-acting form with substitutions at two residues (Figure 5.14). Insulin detemir is a lipid-modified insulin (trade name Levimir) that is longer acting than insulin. Insulin degludec is another long-acting lipidated form.

Amylin, also known as islet amyloid polypeptide (IAPP), is a peptide co-secreted with insulin from pancreatic beta cells. It is involved in the regulation of glycaemia, promoting satiety via gastric emptying. An amylin analogue, pramlintide is used to treat diabetes (both types I and II). Like amylin, it is a 37-residue peptide containing a disulfide bond and acts on hypothalamic receptors. Pramlintide differs from amylin in substitutions of three residues for prolines (it is [25,28,29]Pro-human amylin). As its name suggests, amylin forms amyloid structures *in vitro*, and the peptide and fragments from it have been used as model amyloid peptides (see Section 3.6).

Glucagon is an important pancreatic hormone that serves as the primary catabolic peptide in the body, i.e. it is involved in the metabolic breakdown of biomolecules that generate energy. Glucagon raises the concentration of

Figure 5.15 α-helical structure of glucagon, obtained from X-ray diffraction measurements.

glucose and lipids in the bloodstream. It is a linear 29-residue peptide cleaved from the proglucagon protein, and it has an α-helical structure (Figure 5.15). It can be used therapeutically to treat low blood sugar levels in diabetics; in particular it is injected to treat severe hypoglycaemia. Glucagon is related to several glucagon-like peptides (also cleaved from pro-glucagon). It belongs to the secretin family of peptides (Section 5.3.5).

5.3.4   Posterior Pituitary (Neurohypophysis) Hormones

Oxytocin is a member of a series of related peptide hormones that are non-apeptides and octapeptides; they have an intramolecular disulfide bond that produces a small cyclic ring (Figure 5.16). These peptides are ligands for the G-protein coupled oxytocin receptor (OTR). Oxytocin is used medically to facilitate childbirth (it simulates uterine contractions) and milk secretion. It may also have a role in promoting social bonding. Demoxytocin (or desaminooxytocin or deaminooxytocin) is an analogue of oxytocin in which Cys is replaced with Mpa (β-mercaptopropionic acid). Demoxytocin has longer circulation half-life than oxytocin and has similar applications. Atosiban is a nonapeptide, desamino acid, analogue of oxytocin used to prevent premature labour. It is a vasopressin/OTR agonist. Carbetocin is an OTR agonist used to control postpartum haemorrhage and bleeding and is an octapeptide analogue of oxytocin.

The vasopressins are a family of nine-residue peptide hormones related to the oxytocin family, and which are also released from the posterior pituitary (from a prohormone produced in the hypothalamus). The vasopressins contain a disulfide bond that creates a cyclic region within the molecule. Vasopressins act as vasoconstrictors and are thus targets for blood pressure modulation. They also stimulate water resorption by the kidneys and are thus antidiuretic agents. Arginine vasopressin (Figure 5.16), also known as argipressin or pitressin, is the basis for the antidiuretic (and anticoagulant) desmopressin, which has the arginine vasopressin sequence (Figure 5.16)

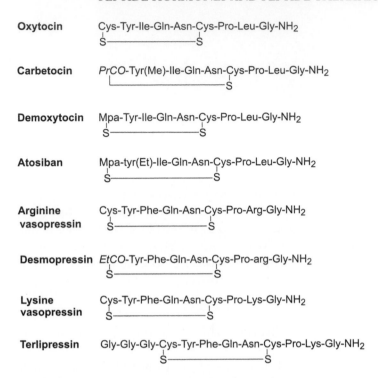

Figure 5.16 Vasopressin/oxytocin peptide family members. Mpa = 3-mercapto-propioic acid (tyr denotes D-Tyr and arg denotes D-Arg). *EtCO* denotes CH_2CH_2CO and *PrCO* denotes $CH_2CH_2CH_2CO$.

but with the first C residue deamidated and with D-Arg instead of L-Arg. Vasopressins act on vasopressin receptors, which are GPCRs. Desmopressin is used as a treatment for diabetes insipidus, bedwetting, and increased thirst or urination associated with head surgery or head trauma. Lysine vasopressin (lypressin) is also a vasopressin-derived peptide (with Lys replacing Arg, Figure 5.16) used to treat diabetes insipidus. Terlipressin is a lysine vasopressin analogue (with Phe substituting for Tyr at position 2) with indications for use including oesophageal varices (dilated veins in the oesophagus), septic shock, and hepatorenal syndrome in kidney failure. Felypressin is a vasopressin analogue added to dental anaesthetics as it is a vasoconstrictor and reduces pain and inflammation. Ornipressin is another vasopressin-derived peptide (with ornithine replacing Arg in the arginine vasopressin sequence) used to reduce bleeding.

The angiotensins are a series of four linear peptide hormones. The first, angiotensin I, is obtained by cleavage of the angiotensinogen protein by the 340-residue protein renin. Angiotensin I has the sequence NRVYIHPFHL. Angiotensin II is derived from this by cleavage of the two C-terminal

residues, angiotensin III is the heptapeptide RVYIHPF, and angiotensin
IV is the core hexapeptide VYIHPF. Angiotensin II has the important
bioactivity of vasoconstriction and blood pressure increase. It is produced
from angiotensin I by angiotensin-converting enzyme (ACE) and hence
ACE inhibitors are therapeutic agents for blood pressure reduction and cardiovascular disease. Angiotensin II receptor antagonists are also important
therapeutic agents. Angiotensin II has an identified role in inflammation and
it can regulate growth factors and cytokines. It was approved by the FDA in
2017 as an active vasoconstrictor (trade name Giapreza) to increase blood
pressure in those suffering from septic shock. Lisinopril (Figure 5.17a) is a
very important peptide-based ACE inhibitor which is very widely prescribed.
It was the second most prescribed drug in the US in 2016 with 110 million
prescriptions. The related compound enalapril (Figure 5.17b) is also on the
WHO (World Health Organization) list of essential medicines. Captopril
(D-2-methyl-3-mercaptopropanoyl-L-proline) is a related ACE inhibitor.

Bradykinin, with the sequence RPPGFSPFR, is a short peptide hormone
with activity in blood vessel dilation (vasodilation), and hence in reduction
of blood pressure. It is also associated with inflammatory pain. The ACE
inhibitors increase bradykinin production. Icatibant is an orphan drug used
in the treatment of hereditary angioedema. It is a 10-residue peptide derivative (with a number of non-natural residues), and acts as an antagonist of
bradykinin B2 receptors. Ecallantide is another peptide treatment for hereditary angioedema. It is a 60-residue peptide containing three disulfide bridges.
Ecallantide is an inhibitor of the kallikrein subgroup of proteases, and is
derived from a Kunitz domain (a protein domain protease inhibitor) from a
tissue factor pathway inhibitor.

Atrial natriuretic peptide (ANP) is a 28-residue peptide containing a
17-residue disulfide bridge-stabilized cycle. It increases sodium and water

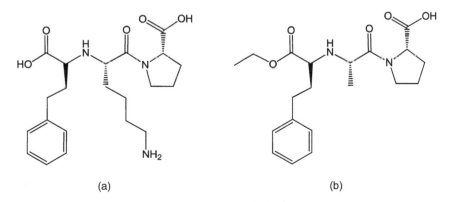

(a)   (b)

Figure 5.17   Molecular structures of (a) lisinopril and (b) enalapril.

excretion (natriuresis) in the kidneys. It is secreted from cardiac muscle cells. ANP is one of a family of nine structurally similar natriuretic hormones: seven are produced in the cardiac atria, including the misnamed brain natriuretic peptide (BNP) and CNP (C-type natriuretic peptide). ANP is also known as cardiodilatin. It causes a reduction in expanded extracellular fluid (ECF) volume by increasing renal sodium excretion. Urodilatin is a related peptide hormone that causes natriuresis via increased renal blood flow. It is longer by four residues at the N terminus compared to ANP and is cleaved from the same precursor. BNP and CNP are discussed further in Section 5.4. ANP also inhibits angiotensin production.

5.3.5   Gut Hormones

Gut–brain interactions are increasingly recognized as playing an important role in determining overall food intake. Many peptides are synthesized and released from the gastrointestinal tract, and it has been shown that they physiologically influence eating behaviour via gut–brain signalling. It has been suggested that gut hormones could be manipulated to regulate energy balance, and, as a result, therapies based on gut peptide hormones could be used in possible treatments for obesity. Figure 5.18 shows some of the main peptide hormones within the gut–brain axis.

There are several families of gut hormone peptides: (i) gastrin/cholecystokinin, (ii) secretin, (iii) pancreatic polypeptide (PP), (iv) others. These are discussed in turn in the following sub-sections.

(i) Gastrin and cholecystokinin

Gastrin stimulates the secretion of gastric acid in the stomach. Gastrin peptide hormones have a common C-terminal motif. There are several variants (some too long to fit within the scope of this review) such as minigastrin or gastrin-14, WLEEEEEAYGWMNF.

Cholecystokinin is an endogenous gut hormone mainly found in the duodenum and jejunum that exists in several molecular forms with differing numbers of amino acids. Examples include CCK-8 and CCK-54 (the number indicates the number of amino acid residues). CCK is known to act as a postprandial satiety signal and it acts via two receptors: CCK_1 and CCK_2. The CCK_1 receptor is more important in appetite control. The receptors are located on the peripheral vagal afferent terminals, which transmit signals to part of the brain stem that is associated with appetite. Cholecystokinin tetrapeptide (CCK-4), also known as tetragastrin, is a peptide

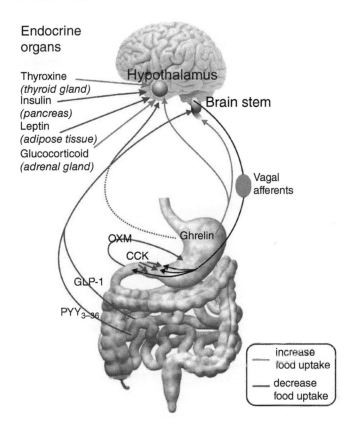

Figure 5.18 Interactions of gut and endocrine hormones with the brain and how they affect food intake. The peptides leptin, ghrelin, OXM (oxyntomodulin), CCK (cholecystokinin), GLP-1 (glucagon-like peptide 1), and PYY_{3-36} are discussed in the text.

fragment derived from the larger peptide hormone, and has the sequence WMNF. It is an anxiolytic and is used to induce panic attacks in studies of anxiety.

(ii) Secretins and related peptides

Secretin itself is a 27-residue peptide produced in the intestine that plays an essential role in water homeostasis throughout the body, that regulates secretions in the stomach, pancreas, and liver, and regulates the pH of the duodenum. Secretin also acts in concert with CCK to release bile salts from the gallbladder.

Glucagon-like peptide 1 (GLP-1) is a 37-amino acid gut peptide which is a proglucagon-based peptide. The sequence is shown in Figure 5.19a. It is

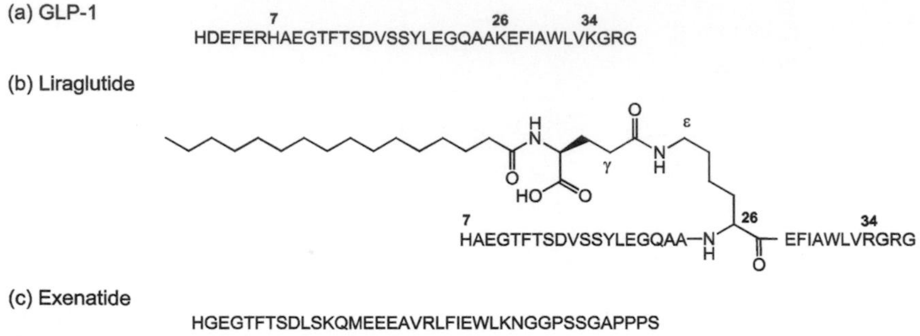

(a) GLP-1

                        7   26   34
HDEFERHAEGTFTSDVSSYLEGQAAKEFIAWLVKGRG

(b) Liraglutide

             7   26   34
HAEGTFTSDVSSYLEGQAA—N   EFIAWLVRGRG

(c) Exenatide

HGEGTFTSDLSKQMEEEAVRLFIEWLKNGGPSSGAPPPS

Figure 5.19 Molecular structures of: (a) GLP-1 with residues which are substituted/derivatized, (b) liraglutide, and (c) exenatide.

an incretin peptide hormone, meaning that it stimulates insulin secretion in response to eating, and as a result it suppresses glucagon secretion. In addition to this, GLP-1 inhibits gastric emptying, and thus reduces food intake due to a sensation of increased satiety. GLP-1 is produced in the intestinal epithelial endocrine L-cells in the distal small bowel and colon by differential processing of proglucagon. Within minutes of food intake, the plasma levels of GLP-1 rise rapidly. GLP-1 exists in two circulating molecular forms: GLP-1(7–37) and GLP-1(7–36) amide, with the latter representing the majority of circulating active GLP-1 in human plasma. Both forms of GLP-1 are rapidly metabolized and inactivated by the enzyme dipeptidyl peptidase-4 (DPP-4) to GLP-1(9–37) or GLP-1(9–36)-NH_2 following release from gut L-cells. This widely expressed enzyme cleaves both forms of GLP-1 after the second N-terminal alanine residue (Ala8) to make them inactive. The expression of DPP-4 in the gut and vascular endothelium explains the short half-life of GLP-1 of just several minutes, because the majority of GLP-1 entering venous circulation has already been inactivated by N-terminal cleavage. Some peptide therapies based on GLP-1 involve substitutions to suppress DPP-4 cleavage, as well as other substitutions and lipidation to improve circulation time.

Liraglutide is a peptide hormone developed to treat type II diabetes and obesity. Liraglutide is a GLP-1 receptor agonist. Lipid conjugation of a palmitoyl chain to a lysine residue at position 26 within the GLP-1(7–27) sequence (Figure 5.19b) results in an extended half-life (around 13–14 hours) in the blood. The palmitoyl chain allows non-covalent binding to albumin, which delays proteolytic attack by DPP-4 and renal clearance. Furthermore, the addition of the lipid chain can also prolong half-life by sterically hindering the DPP-4 enzyme from degradation. Semaglutide has a related structure and application. It has a longer stearic acid chain linked via a diacid-based spacer at lysine 26 in GLP-1(7–27). It has been developed as the first oral GLP-1 treatment (Rybelsus) for type II diabetes and is longer acting than liraglutide.

Another GLP-1 receptor agonist with similar applications in the treatment of type II diabetes is exenatide (trade names Byetta or Bydureon). The molecular structure is shown in Figure 5.19c. Exenatide is a 39-residue peptide, is derived from a Gila monster toxin peptide, and has a longer half-life *in vivo* than GLP-1. Other GLP-1 agonists that have been marketed as treatments for type II diabetes include semaglutide, dulaglutide, and albiglutide. Semaglutide is a GLP-1(7–37)-based peptide with two amino acid substitutions at positions 8 and 34, where alanine and lysine are replaced by 2-aminoisobutyric acid and arginine respectively. Dulaglutide comprises a GLP-1(7–37) sequence (with two substitutions) linked to an immunoglobulin G antibody tail sequence. This class of molecule is known as a peptibody. Albiglutide comprises a similar GLP-1-based sequence conjugated to human albumin. Lixisenatide (Adlyxin) is a longer GLP-1 agonist peptide, which, like exenatide, is derived from Gila monster exendin 4. Other venom-based peptide therapeutics are discussed in Section 5.5. Many other GLP-1 receptor agonist peptides are in development at various stages of clinical trials, as this is one of the key areas for peptide therapeutics.

Like GLP-1, GLP-2 is also obtained from proglucagon in the intestine, although it is not such an important target as GLP-1. GLP-2 receptors have also been targeted in the development of peptide therapeutics. Teduglutide is a GLP-2 analogue orphan drug developed to treat short bowel syndrome. It differs from the GLP-2 peptide in only one residue substitution, which reduces degradation by DPP-4 and extends circulation time.

Gastric inhibitory peptide (GIP) is a 43-residue peptide which, as its name suggests, inhibits gastric acid section and emptying. It is produced in the duodenum and its major role is to stimulate pancreatic insulin release.

Vasoactive intestinal peptide (VIP) is a 28-residue peptide (HSDAVFTD-NYTRLRKQMAVKKYLNSILN) that causes vasodilation and stimulates heart contraction. It is produced in the gut, pancreas, and suprachiasmatic nuclei of the hypothalamus. It belongs to the glucagon/secretin superfamily. Aviptadil has the same sequence as VIP and is used to treat erectile dysfunction.

Oxyntomodulin (OXM) is a 37-amino acid peptide that is derived from the proglucagon gene. It acts on G-protein coupled receptors of the secretin family (including the GLP-1 and GLP-2 receptors), the glucagon receptor, and the GIP receptor. The first N-terminal sequence is a 29-residue sequence from glucagon with an eight-residue extension at the C-terminus. Oxyntomodulin mediates its effects via the GLP-1 receptor, as shown in experiments carried out with rat parietal cells. Its anorectic (appetite suppression) actions are blocked when a GLP-1 antagonist is administered. Intravenous administration of OXM in humans inhibits gastric emptying and gastric acid secretion, which leads to a feeling of satiety. This feeling of satiety can cause a reduction in both food intake and overall body weight, and this is brought about by the suppression of ghrelin (discussed shortly).

(iii) Pancreatic polypeptides

The pancreatic polypeptide (PP) family includes PP itself, as well as peptide YY (PYY) and neuropeptide Y (NPY), all of which contain the PP peptide fold. The sequences of the human versions of these peptides are listed in Table 5.3.

The PP fold involves certain common residues which are predominately Pro2, Pro5, Pro8, Gly9, Tyr20, and Tyr27. This PP fold has been found to protect the peptide against enzymatic attack and the hydrophobic pocket produces a favourable folded structure.

The PP fold pancreatic polypeptide family all consist of a signal peptide, followed by a 36-amino acid active peptide and a carboxyl terminus. They mediate their effects through binding to the Y receptors Y_1–Y_6. The Y receptors belong to the G protein-coupled receptor family and they mediate a wide variety of physiological effects, such as regulation of blood pressure, anxiety, memory retention, hormone release, and food intake.

PP is a 36-residue peptide with high affinity for the Y_4 and Y_5 receptors. PP is similar to GLP-1 in that it is released into the circulation after the ingestion of food. However, it differs in that it is produced in the endocrine F cells, which are located in the periphery of pancreatic islets. PP is responsible for a number of regulatory actions, such as the inhibition of pancreatic exocrine secretion, the modulation of gastric acid secretion, and gastric emptying. The amount of PP released is affected by the digestive state, i.e. release is very low in the fasted state, but is significantly increased throughout all phases of digestion. In addition to this, PP is affected by a decrease in blood

Table 5.3   Sequences of human pancreatic polypeptides with PP-fold related residues in bold.

Peptide	Sequence
PP	A P L E **P** V Y **P G** D N A T P E Q M A Q **Y** A A D L R R **Y** I N M L T R P R Y
PYY	Y **P** I K **P** E A **P G** E D A S P E E L N R **Y** Y A S L R H **Y** L N L V T R Q R Y
NPY	Y **P** S K **P** D N **P G** E D A P A E D M A R **Y** Y S A L R H **Y** I N L I T R Q R Y

glucose levels and insulin-induced hypoglycaemia in that they are stimuli for PP secretion in the brain. As a result of this, it is thought that PP could potentially play a significant role in the regulation of feeding behaviour to control energy homeostasis.

In addition to containing the PP fold motif, PYY and its truncated derivative PYY_{3-36} have a high C-terminal α-helix content, which has also been suggested to be extremely important for the structural integrity of PYY. PYY is released by the L-cells of the gastrointestinal tract following food intake, and there are two main endogenous forms: PYY and PYY_{3-36}. The full PYY peptide, PYY_{1-36}, is rapidly processed by the enzyme DPP-4 to the 34-amino acid peptide PYY_{3-36}. DPP-4 hydrolyses PYY and removes the first two amino acids, tyrosine and proline, at the N-terminal, which changes the receptor selectivity. As a result of this, PYY_{3-36} has a high selectivity for the Y_2 receptor, compared to PYY which has selectivity for the Y_1, Y_2, and Y_5 receptors. It is thought that the Y_1 receptor requires both the C-terminus and N-terminus for recognition, binding, and then subsequent activation. The Y_2 receptor is thought to have a smaller receptor site and also only requires the C-terminus for recognition. This leads to the reduced affinity for PYY_{3-36} on any Y receptor other than Y_2. The Y_2 receptors are located in the hippocampus, sympathetic and parasympathetic nerve fibres, intestines, and certain blood vessels, and have been implicated in regulating food intake and gastric emptying. As a result of this, the Y_2 receptor is considered a target for the treatment of obesity and type 2 diabetes.

Neuropeptide Y (NPY) is also a 36-amino acid peptide that has a very similar sequence homology to peptide YY (PYY) and PP (Table 5.3). NPY differs from the other two peptides because it acts as a neurotransmitter. NPY is one of the most abundant peptides found in the brain and it is synthesized and released by neurons. NPY is associated with various biological responses, including increased food intake, enhanced cognitive function associated with learning and memory, and also reduction in anxiety. In addition to this, NPY has been shown to induce vasoconstriction in peripheral blood vessels. Studies of NPY and its receptors suggest that it could be directly related to various pathological disorders such as obesity, depression, and epilepsy.

(iv) Other gut hormones

Ghrelin is a 28-amino acid peptide with an octanoylated serine residue at position 3; it is produced and secreted by cells within the oxyntic glands of the stomach. Peripheral administration of ghrelin has been shown to stimulate food intake and decrease fat utilization. Ghrelin and galanin (discussed in Section 5.3.6) increase food (especially fat) intake and as such are targets

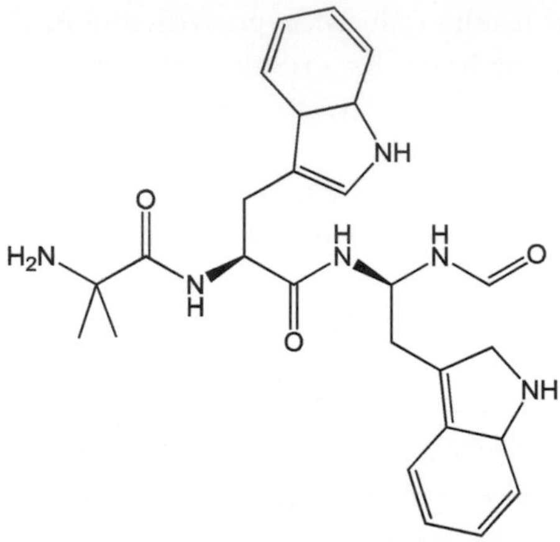

Figure 5.20   Molecular structure of macimorelin.

for weight control therapies. Ghrelin provided the basis for the development of the growth hormone secretogogue (a substance that promotes secretion) macimorelin; macimorelin is a pseudopeptide with molecular structure shown in Figure 5.20. Macimorelin can be administered orally in the treatment of adult growth hormone deficiency.

Motilin is a 22-residue peptide involved in the regulation of contractions of the intestine. It also stimulates endogenous release of hormones from the pancreas.

5.3.6   Neurotransmitters Produced in the Gut

Table 5.4 lists neurotransmitter peptides associated with the gastrointestinal tract as part of the gut–brain axis.

Gastrin-releasing peptide (GRP) is produced by a gene that encodes a number of bombesin-like peptides. It is a 27-residue neuropeptide with a role, as its name suggests, in stimulating gastrin release in the stomach but it also has a role in the development of the lung epithelium. GRP shares a common C-terminal 10-residue sequence homology with bombesin. Bombesin is a 14-residue peptide originally isolated from toad skin. It is the prototype member of a family of bombesin-related peptides that include GRP and neuromedins B and C. Bombesin peptides secreted by amphibians probably have a defensive function; however, they have a range of activities in mammals, including stimulation of gastrin release in the stomach. Bombesin-related peptides are recognized by the gastrin-releasing-peptide receptor (GRP-R), which is overexpressed on several types of cancer cell including prostate,

Table 5.4 Gut-related peptide families including neurotransmitters (italicized).

Family	Members
Gastrin-releasing peptide	*Gastrin-releasing peptide (GRP), neuromedin B, neuromedin C,* bombesin, ranotensin, litorin
Calcitonin gene-related peptide	*Calcitonin gene-related peptide (CGRP),* adrenomedullin, amylin, calcitonin
Neurotensins	*Neurotensin, neuromedin N*
Tachykinins	*Substance P, neurokinin A, neurokinin B, neuromedin K, neuromedin L,* eledoisin
Pancreatic polypeptide	*Neuropeptide Y (NPY),* pancreatic polypeptide (PP), peptide YY (PYY)
Others	*Galanin*

breast, and small cell lung cancer cells. These peptides are therefore interesting candidates in the development of cancer therapeutics.

Neuromedin B is a decapeptide member of the bombesin family, and, in humans, is expressed in the CNS, peripheral organs, and gastrointestinal tract. Neuromedin C is a related decapeptide with strong sequence homology to the other members of the bombesin peptide family. Bombesin has the sequence (pGlu)QRLGNQWAVGHIKM-NH$_2$ (pGlu is pyroglutamic acid, see Table 1.3), human GRP is VPLPAGGGTVLTKMYPRGN-HWAVGHLM, human neuromedin B has the sequence NLWATGHFM, and human neuromedin C is GNHWAVGHLM-NH$_2$. These peptides have a common C-terminal sequence WAxGHyM (x = T, V, y = K, L, F). This sequence is shared with the phyllotorins and ranotensin and litorin amphibian skin-derived peptides.

Calcitonin gene-related peptide (CGRP) is a member of the CGRP family that includes calcitonin, adrenomedullin, and amylin (discussed in Section 5.3.3). CGRP is a 37-residue peptide produced by neurons. It is a potent peptide vasodilator and has a role in signalling in the nervous system. Adrenomedullin is a 52-amino acid peptide (with one disulfide bridge) and also acts as a vasodilator.

Neurotensin is a 13-residue peptide found throughout the CNS. It is involved in the regulation of the release of luteinizing hormone and prolactin (a protein enabling milk production) release and has significant interaction with the dopaminergic system. It is believed to have a variety of roles *in vivo* including gastrointestinal (GI) motility, glucose homeostasis, analgesia, hypothermia, increased locomotor activity, and cell proliferation. Human neurotensin with sequence QLYENKPRRPYIL has a C-terminus in common with another neuropeptide neuromedin N (IPYIL), which has similar roles.

Substance P is the undecapeptide RPKPQQFFGLM; it acts as a neurotransmitter and is involved in vasodilation, inflammation, pain, and analgesia, among other biological activities. It is a member of the tachykinin family of neuropeptides, which have the common C-terminus FxGLM where x is a hydrophobic residue. Neurokinin A is involved in the mammalian neuroinflammatory and pain response. Neurokinin B is involved in the signalling pathway for the secretion of GnRH and has roles in regulation of the ovarian cycle, pregnancy, and postmenopausal symptoms. Neuromedins K and L also belong to the tachykinin family. Eledoisin A is a 10-residue peptide member of the tachykinin family that is derived from mollusc salivary glands. It has similar activity to substance P although it was originally developed as a treatment for dry eyes, as it is a stimulant of lachrymal secretion.

NPY is discussed along with the other pancreatic polypeptides in Section 5.3.5; however, it also acts as a neurotransmitter.

Galanin is a 29- or 30-residue peptide found in the gastrointestinal tract and CNS and it has a variety of roles, including nervous response to stimulus (nociception), appetite stimulation, and regulation of blood pressure.

5.4   NEUROPEPTIDES AND OTHER PEPTIDES
IN VIVO

Neurotransmitter peptides associated with the essential gut–brain axis are discussed in Section 5.3.6. The remaining peptides in this section are a miscellany of peptides expressed in other parts of the body, with various functions.

Leptin is a peptide hormone made by adipose cells that affects many biological mechanisms, including reproduction, the immune and inflammatory response, haematopoiesis (production of blood), angiogenesis, bone formation, and wound healing. More interestingly, however, leptin helps to regulate energy balance by inhibiting hunger. This occurs via a feedback mechanism in which signals are sent to key regulatory centres in the brain to inhibit food intake (see Figure 5.18). After leptin is released by adipose tissue into the bloodstream, it crosses the BBB and binds to the hypothalamic leptin receptors. This affects the activity of hypothalamic neurons, and the expression of various orexigenic (appetite stimulating) and anorexigenic neuropeptides. Orexigenic peptides include NPY, and anorexigenics include PP and POMC. It has been suggested that the interaction with both types of these neuropeptides underpins the mechanism of action of leptin in the hypothalamus in reducing hunger.

Brain natriuretic peptide (BNP), mentioned in Section 5.3.4, is secreted by cardiomyocytes in heart ventricles. It is involved in the regulation of blood pressure via natriuresis (sodium excretion in urine) as well as vascular resistance and systemic blood pressure. It is a 32-residue peptide with a single disulfide bridge. Nesiritide, a recombinant form of BNP, was proposed as a treatment for acute decompensated heart failure although there is currently controversy about its efficacy.

C-type natriuretic peptide (CNP) is cleaved from pro-CNP into three peptides with 22, 29, and 53 residues. CNP is produced in the endothelium in response to stress and proinflammatory stimuli. It does not have direct natriuretic activity.

Endolithins are 21-residue peptides which act as vasoconstrictors. They are produced by endothelial cells (in particular, those on the surface of blood and lymphatic vessels). They are also believed to have a role in cardiovascular function and in electrolyte regulation in the body.

The enkephalins are natural opioid receptor peptides. There are two enkephalin peptides: Met-enkephalin (YGGFM) and Leu-enkephalin (YGGFL), which are endogenous ligands for μ-opioid receptors in the brain. Dalargin, with sequence YaGFLR (a = Δ-alanine) is a Leu-enkephalin analogue with specificity for μ-opioid receptors. Dalargin is usually excluded from the brain; however, lipidation leads to a self-assembled fibril structure that is reported to both reduce degradation in plasma and enable transport across the BBB. The endomorphin tetrapeptides also have high selectivity and affinity for the μ-opioid receptors. Two endomorphin peptides were isolated from brain extracts. Endomorphin-1 has the sequence YPWF and endomorphin-2 has the sequence YPFF. These peptides are of interest for the further development of pain-relief therapies. However, it has to be noted that the direct application of these peptides in analgesia was demonstrated in a rodent model by i.c.v. (intracerebroventricular) administration. Peripheral administration of these peptides is ineffective as they are too easily degraded; appropriately modified versions may be designed to circumvent this.

Glutathione is tripeptide (L-γ-glutamyl-L-cysteinylglycine), and is denoted GSH, although there is a gamma peptide link between the glutamic acid side chain and cysteine. It is an antioxidant, the thiol group serving as a reducing agent *in vivo*, disrupting disulfide bonds in proteins to yield free cysteine groups. It is involved in many biological processes, including metabolism, protein and DNA synthesis, and protection of cells against oxidation. Glutathione can be produced by all animal cells; however, glutathione synthesis in the liver has been shown to be essential.

Thymopoietin (also known as lamina-associated polypeptide 2) is a protein in the thymus, a gland which produces T-cells that are the key component of the adaptive immune system. Thymopentin, also known as TP-5, is a pentapeptide fragment (RLNVY) corresponding to residues 32 to 36 of human thymopoietin. Thymopentin is an immunostimulant used, for example, in the treatment of dermatitis among many other medical applications involving immunomodulation. Splenopentin, sequence RLGVY, is a related immunomodulation peptide.

Thymosins were originally isolated from the thymus; however, they are now known to be present in other tissues. Two variants, thymosin-β_4 and thymosin-α_1, have clinical relevance. Thymosin-β_4 is a 43-residue peptide that is being investigated for applications as an anti-inflammatory and in tissue repair. Thymosin-α_1 is used as a treatment for hepatitis B and hepatitis C (drug tradename Zadaxin) and to boost immune response in the treatment of other conditions, for example as an adjuvant in cancer treatments. It is a 28-residue peptide derived from a longer precursor protein.

5.5 VENOM-DERIVED PEPTIDES

Many animals, such as amphibians, scorpions, snakes, spiders, bees, wasps, jellyfish, sea urchins, cone snails, and others, have evolved venoms as part of their defence or predation strategies. Many venoms are peptides that have evolved to target cells with considerable specificity and potency. They have thus attracted attention as scaffolds to produce new drugs. Indeed, several peptide therapeutics on the market have been derived in this fashion, as listed in Table 5.5.

Exenatide, since it is derived from a peptide hormone, is discussed in Section 5.3.5.

Eptifibatide (trade name Integrillin) is an antiplatelet drug used to reduce the risk of acute cardiac ischemic events such as heart attack. It is a cyclic heptapeptide (shown in Figure 5.21a) that contains an RGD mimetic sequence Mpr-HomoArg-GDWPC-NH$_2$ (cyclized via a mercaptopropionic acid, Mpr, bridge) (HomoArg is homoarginine, see Table 1.3). It was derived from a disintegrin protein found in the venom of the southeastern pygmy rattlesnake. Tirofiban is another antiplatelet drug derived from a snake venom, but as it is not a peptide itself it is not discussed further here.

Bivalirudin (trade name Angiomax or Angiox) is a direct thrombin inhibitor and is used as an anticoagulant for patients undergoing coronary interventions. It is a 20-amino acid peptide with structure shown in Figure 5.21b (sequence fPRPGGGGMGDFEEIPEEYL). It was derived from the saliva of leeches, used traditionally in blood letting. The leech saliva contains a large 65-residue peptide hirudin that was further developed as

Table 5.5   Venom-derived peptide therapeutics.

Peptide	Origin	Activity	Application
Eptifibatide	Disintegrin protein from pygmy rattlesnake	Integrin receptor	Acute coronary conditions
Exenatide	Synthetic version of exendin-4, a hormone found in the saliva of the Gila monster	GLP-1 agonist	Type II diabetes
Bivalirudin	Leech saliva mini-protein hirudin	Thrombin inhibitor	Blood coagulation during surgery
Ziconotide	ω-conotoxin peptide from cone snail	Calcium ion channel blocker	Chronic pain
Linaclotide	*E. coli* enterotoxin	Agonist of large intestine GPCR 2C	Irritable bowel syndrome and constipation
Plecanatide	*E. coli* enterotoxin	Agonist of large intestine GPCR 2C	Irritable bowel syndrome and constipation

the injectable drug bivalirudin. Lepirudin (trade name Refludan) is a related antithromobolytic based closely on the original hirudin mini-protein, having only two minor substitutions/modifications.

Ziconotide is a powerful analgesic that was derived from a ω-conotoxin obtained from the cone shell *Conus magus*. Cone shells are highly dangerous (potentially lethal) organisms that deliver their toxins via harpoon injection. These venoms act on voltage-activated N-type calcium ion channels that are involved in neurotransmitter release. This suggested a strategy to develop an analgesic based on inhibition of N-type calcium channels, which led to ziconotide. The structure, which is stabilized by three disulfide bonds, is shown in Figure 5.22a.

Linaclotide (Figure 5.22b) came to market in 2012 as an oral peptide therapeutic to treat irritable bowel syndrome and chronic constipation. The drug acts on GPCRs in the large intestine. It was designed based on *Escherichia coli* bacterial enterotoxins, which are highly heat stable. These peptides are closely related to uroguanylins, occurring naturally in animals, that regulate water and electrolytes in the intestine and kidneys. The three disulfide bridges (Figure 5.22b) provide stability to the molecule. Plecanatide (Figure 5.22c) was derived from the same source, and also stimulates water release into the gastrointestinal tract, softening stools and so relieving chronic constipation.

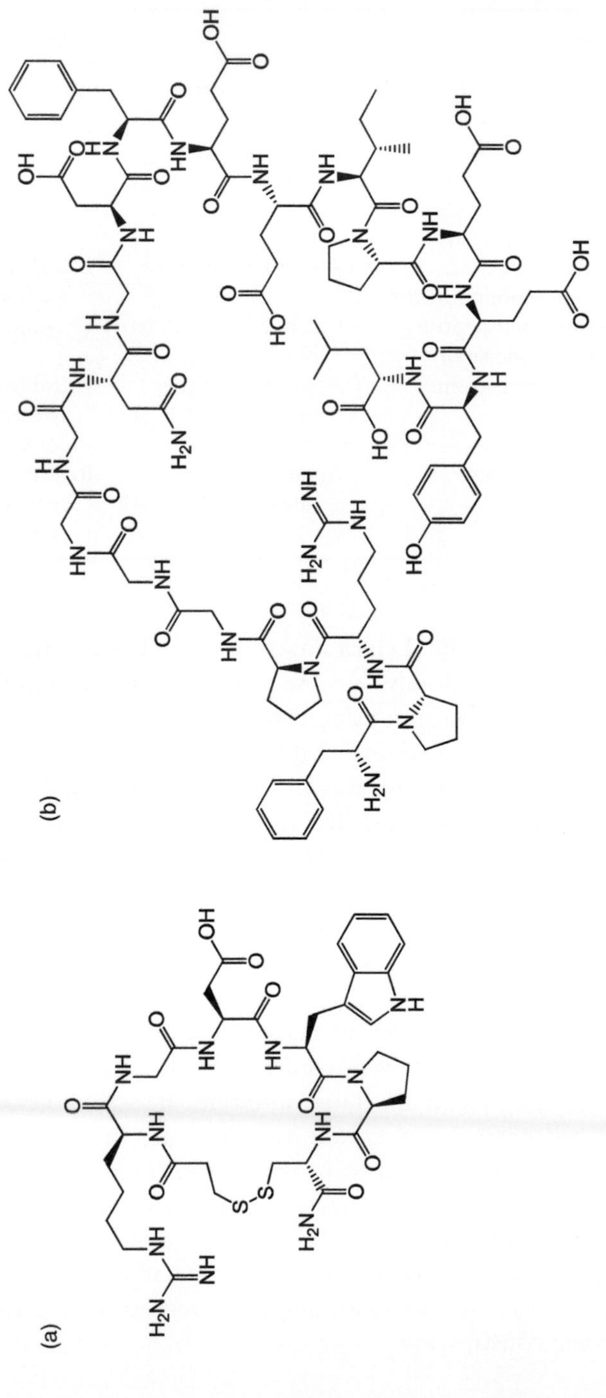

Figure 5.21 Venom-based peptide therapeutics: (a) eptifibatide and (b) bivalirudin.

(a)

Cys-Lys-Gly-Lys-Gly-Ala-Lys-Cys-
Ser-Arg-Leu-Met-Tyr-Asp-Cys-Cys-
Thr-Gly-Ser-Cys-Arg-Ser-Lys-Cys-NH$_2$

(b)
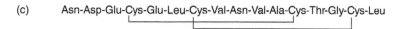
Cys-Cys-Glu-Tyr-Cys-Cys-Asn-Pro-Ala-Cys-Thr-Gly-Cys-Tyr

(c)
Asn-Asp-Glu-Cys-Glu-Leu-Cys-Val-Asn-Val-Ala-Cys-Thr-Gly-Cys-Leu

Figure 5.22 Molecular structures of (a) ziconotide, (b) linaclotide, and (c) plecanatide.

5.6   ANTICANCER PEPTIDES

There are many pathways that can be targeted in the development of anti-cancer agents. This is reflected in the diversity of models of action of anti-cancer peptides that are described in this section.

Several GnRH-agonist peptides with anticancer activity have been discussed above and are listed in Table 5.2. Abarelix and degarelix, also discussed in Section 5.3.1, are treatments for prostate cancer. GHRH (ant)agonists (also discussed in Section 5.3.1) are important peptide treatments for several cancers.

ADH-1 is a cyclic disulfide-bridged pentapeptide derivative developed to treat melanoma and is under investigation for other cancer treatments. It blocks the action of N-cadherin, which plays a role in cancer metastasis. ADH-1 can disrupt tumour cell development and tumour vasculature.

Bortezomib is an N-protected dipeptide used as a chemotherapy agent to treat multiple myeloma and mantle cell lymphoma. It has the structure shown in Figure 5.23a, i.e. where Pyz-Phe-Leu-BOOH, Pyz is pyrazinoic acid, with a boronic acid C-terminus. The boron atom prevents degradation by proteasomes and inhibits enzymes including lipases, proteases, and thioesterases, the latter being a cancer drug target (via Ras signalling proteins involved in cell growth).

Apicidin (Figure 5.23b) is a cyclic tetrapeptide that is a histone deacetylase (HDAC) inhibitor, originally isolated as a fungal metabolite. It has potential anticancer activity and acts against protozoan parasites such as *Plasmodium* (the malaria infective agent), among other applications. Chlamydocin is a related cyclic tetrapeptide with origins as a fungal natural product. It is reported to inhibit the growth of tumour cells. Romidepsin (Figure 5.23c) is a bicyclic depsipeptide (isolated from a soil bacterium) which is also a

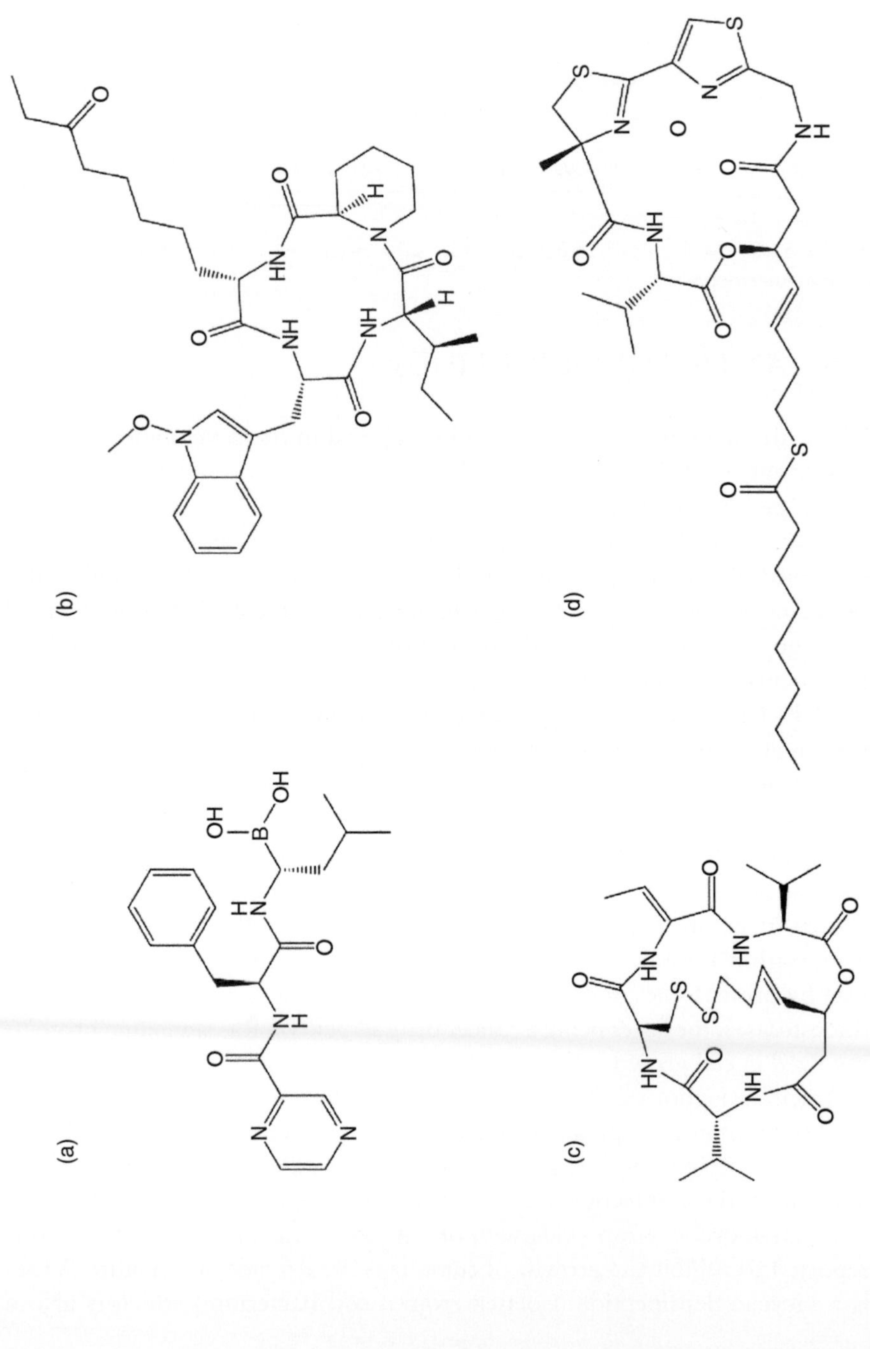

Figure 5.23 Molecular structures of (a) bortezomib, (b) apicidin, (c) romidepsin, and (d) largazole.

HDAC inhibitor with pronounced anticancer activity against several types of lymphoma. Largazole (Figure 5.23d) also belongs to the class of cyclic depsipeptide natural products with HDAC anticancer activity. A variety of other, less well-studied, cyclic peptides with anticancer activity are known.

Several anticancer peptides were originally isolated as antibiotics. Dactinomycin (actinomycin D) is a peptide derivative used in chemotherapy. It comprises a phenoxazone chromophore with two attached cyclic peptide derivatives (Figure 5.24a). It is an antibiotic derived from a soil bacterium, but is now used due to its anticancer properties. It is believed to interfere with DNA synthesis. Bleomycins are another important class of anticancer agents. The molecules are peptide derivatives (Figure 5.24b), being glycosylated peptide-polyketides containing two thiazolines resulting from cyclization of cysteine residues. The bleomycins are non-ribosomal molecules produced by the bacterium *Streptomyces verticillus*. They act by cleaving DNA strands. Bleomycins have been classed as 'metalloantibiotics' because they were originally discovered as antibiotics and their activity (binding and cleaving of DNA molecules) depends on the presence of ions (especially Fe^{2+}).

Dolastatin 10 (Figure 5.24c) is a short peptide, containing five residues, and with an interesting selection of non-proteogenic amino acids. This peptide, originally isolated from a marine mollusc, has potential anticancer activity, as it has antimitotic activity, inhibiting tubulin polymerization. Dolastatin 10 is one example from a diverse range of marine natural products with anticancer properties resulting from microtubule growth inhibition. Another example is the hemiasterlins, which are short linear peptide derivatives obtained from a marine sponge.

Carfilzomib is a tetrapeptide epoxide used to treat multiple myeloma. It is based on epoxomicin, which is a naturally occurring proteasome inhibitor (a proteasome is a protein complex which facilitates proteolysis). Mifamurtide is a peptide-based drug used to treat osteosarcoma. The structure, shown in Figure 5.25, is based on muramyl dipeptide 2 and it includes a linkage via the phosphoryl group to phosphatidylethanolamine. Muramyl dipeptide 2 is a small immune-stimulating molecule from bacteria, including mycobacteria.

As mentioned above, oncology is a leading area where peptides are the subject of ongoing clinical trials. These have been developed based on a number of strategies.

One strategy is to target integrin receptors as these are involved in angiogenesis and hence tumour progression. As mentioned in Section 3.9, RGD is an integrin α_v receptor-targeting peptide. Cyclic RGD peptides such as cyclo(RGDfV) (f is D-phenylalanine) have superior activity and selectivity compared to linear analogues. This peptide was identified

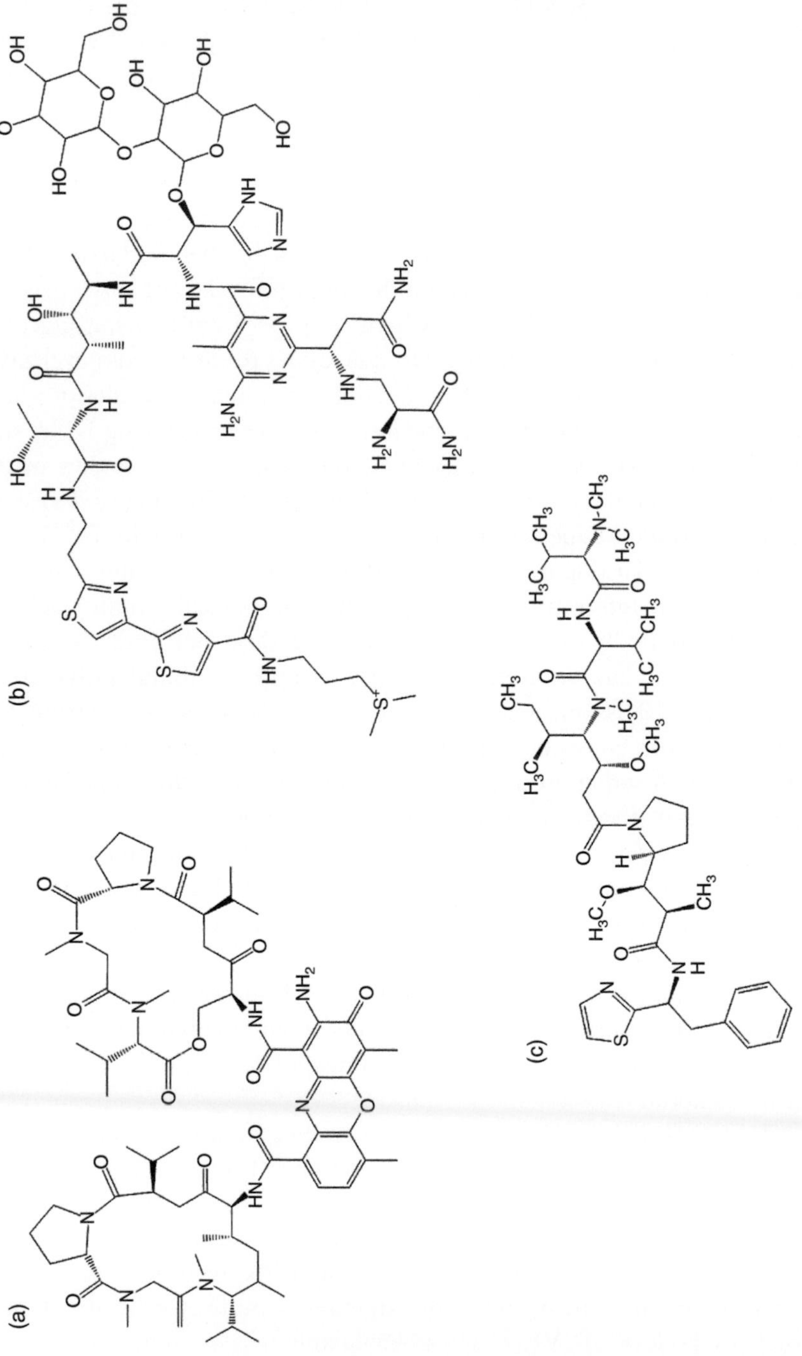

Figure 5.24   Molecular structures of (a) dactinomycin, (b) bleomycin A_2, and (c) dolastatin 10.

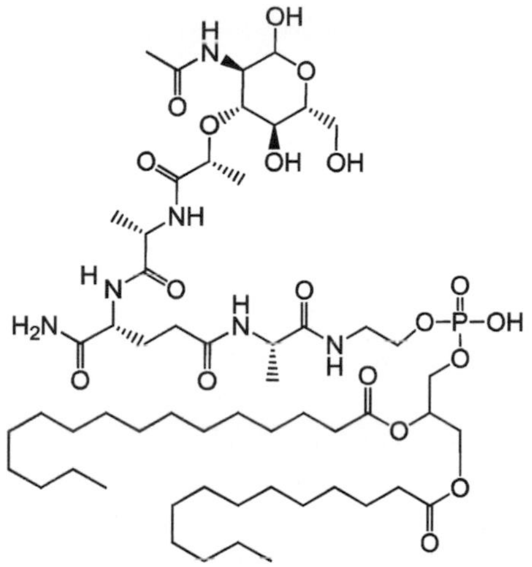

Figure 5.25   Molecular structure of mifamurtide.

through competitive $\alpha_v\beta_3$ integrin cell adhesion inhibition studies. The cyclic N-methylated compound cyclo(RGDf[N-Me]V), with trade name Cilengitide, is under investigation for the treatment of the brain cancer glioblastoma.

Another approach to the development of peptide anticancer actives is to target them as inhibitors of protein kinases. Peptides have potentially useful properties as natural molecules that can block kinases. A related strategy targets phosphatases that are overexpressed in certain kinds of prostate cancer cells. A different approach is to use peptides that are able to penetrate cell membranes, such as antimicrobial and cell-penetrating peptides. These can be used as carriers for peptides targeting particular cancer-related cellular processes or conventional low molecular weight cancer drugs (for example doxorubicin). Some antimicrobial cell-penetrating peptides such as magainin 2 (see Table 4.4) have intrinsic anticancer activity in animal models. A related concept is to target the acidic microenvironment of tumours, as with the pHLIP peptide (AEQNPIYWARYADWLFTTPLLLLDLALLV-DADEGT) and variants which fold into a helical structure and insert across the lipid membrane in response to the low pH (pH ~6) within tumours. Other researchers are exploring the use of enzymatic assembly of peptides (see Section 3.9) within cells to kill them.

Antimicrobial peptide agents are discussed in Chapter 4. A number of these have been shown to have anticancer activity as well. Examples include HNP-1, lactoferricin B, magainin 2, and LL-37 (discussed in Chapter 4). These agents work by targeting cancer cell membranes, entering cells to induce apoptosis.

5.7 MISCELLANEOUS PEPTIDE THERAPEUTICS

Some dipeptides, and even some amino acid derivatives, have biological activities. The amino acid derivative N-acetyl-L-cysteine (NAC), mentioned in Section 5.2, has activity as an enzyme inhibitor, but is also used to treat paracetamol overdose. As it has mucolytic properties, it is used in the treatment of diseases with problematic mucus build-up, including cystic fibrosis and chronic obstructive pulmonary disease. Aminocaproic acid (ε-Ahx, or 6-aminohexanoic acid, see Table 1.3) is a lysine derivative that can be used as an inhibitor for enzymes that are active via lysine residues, including the proteolytic enzyme plasmin. It is thus used to stem acute bleeding due to fibrinolysis (i.e. degradation of fibrinogen, which is a blood-clotting protein).

The dipeptide carnosine, β-AlaHis (β-Ala is β-alanine, see Table 1.3), has a variety of bioactivities including antioxidant and antiglycation properties (relevant to degenerative diseases). It is found in many tissues, especially muscles and the brain. Dipeptide AspPhe-OMe, although not a therapeutic, is the sweetener Aspartame, usage of which should be minimized for those suffering from phenylketonuria (a disease in which phenylalanine metabolism is disrupted).

Cyclosporin, also known as ciclosporin or cyclosporine, is a cyclic 11-residue peptide/peptoid hybrid with sequence cyclo[AbuSar(N(Me)L) V(N(Me)L)Aa(N(Me)L)(N(Me)L)(N(Me)V)-(N(Me)2-Bmt)] (Figure 5.26). Here Sar denotes sarcosine (N-methylglycine), Abu denotes aminobutyric acid (see Table 1.3), and 2-Bmt denotes (E)-2-butenyl-4-methylthreonine. Cyclosporin is a powerful immunosuppressant, active against T-lymphocytes. It also displays antifungal and anti-inflammatory properties. It is an extract from soil fungus and is synthesized by a non-ribosomal synthetase.

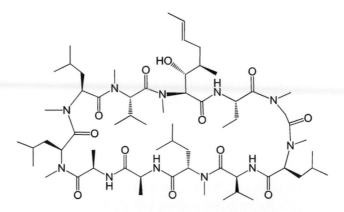

Figure 5.26 Molecular structure of cyclosporin.

Glatiramer (tradename Copaxone) is a treatment for multiple sclerosis (MS) and is a mixture of peptides containing the amino acids glutamic acid, lysine, alanine, and tyrosine that are found in myelin sheath protein (the myelin sheath surrounds nerve cell axons). This protein is believed to stimulate the autoimmune response of patients with MS and glatiramer blocks this, serving as a decoy for the immune cells.

Peptide P-15 was introduced as an FDA-approved material for orthopaedics. It promotes bone grafting and is a collagen-derived peptide GTPGPQGIAGQRGVV, which is a sequence from the α_1 chain of type I collagen that contains a substrate for matrix metalloproteases (enzymes involved in restructuring of the extracellular matrix). This peptide can adsorb onto calcium phosphate during the regrowth of bone.

Enfuvirtide (tradename Fuzeon) is a HIV-fusion inhibitor and is used as an antiretroviral drug in the treatment of HIV. It inhibits the fusion of the gp41 HIV viral coat protein with cell membranes, preventing the virus from entering the cell. It is a 36-residue linear peptide, derived from a naturally occurring motif within a domain of the gp41 transmembrane glycoprotein. The gp41 protein has also been used as a basis to create cell-penetrating peptides, see Table 4.8.

Lucinactant (tradename Surfaxin) is a formulation containing a novel 21-residue repeating peptide sinapultide along with lipids. Sinapultide is KLLLLKLLLLKLLLLKLLLLK, which is designed to mimic human surfactant protein B; human surfactant protein B is a pulmonary surfactant. Lucinactant is used to treat infant respiratory distress syndrome.

Toll-like receptor (TLR) agonist lipopeptides have been developed to stimulate immunogenicity, for example in vaccine adjuvants (substances which modulate immune response). TLRs are transmembrane proteins with a key role in the innate immune system (and also in developing the adaptive immune response) and as such are interesting therapeutic targets. They are activated by Gram-negative bacterial lipopolysaccharide (LPS), an endotoxin. A lipoprotein component of the cell wall of E. coli was identified in 1975 and this led to the development of synthetic TLR2 agonist lipopeptides shown in Figure 5.27 that contain a common CSKKKK peptide motif and one, two, or three ($n = 1, 2, 3$) palmitoyl chains. These peptides are denoted Pam_nCSK_4. These lipopeptides self-assemble into distinct nanostructures, depending on the number of lipid chains. Other TLR agonist lipopeptides have also been studied.

Pam_nCys motifs have been incorporated into so-called self-adjuvating vaccine constructs, by conjugating them to, for example, a helper T-cell epitope peptide sequence and a N-acetylgalactosamine (GalNAc) antigen.

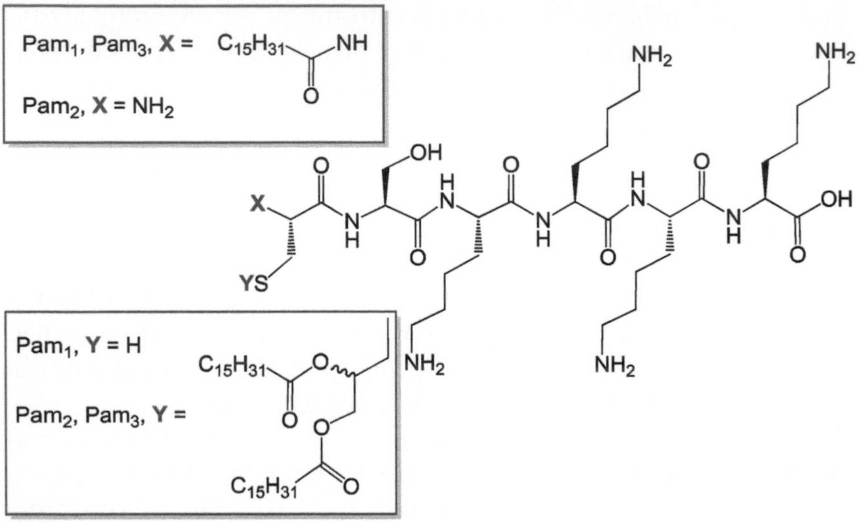

Figure 5.27 Pam_nCSK_4 lipopeptides with $n = 1$, 2, or 3 palmitoyl chains attached to the N-terminal and/or via glycerol linkers to the cysteine residue.

Protein–protein interactions (PPIs) are attracting interest as targets for therapeutics, especially in cancer but also in diseases related to protein aggregation (such as the amyloid diseases discussed in Chapter 3). Peptides (and peptidomimetics) have great potential as PPI inhibitors since they can be designed to present motifs (i.e. amino acids and sequences) that inhibit binding to proteins.

5.8 COSMETIC PEPTIDES AND LIPOPEPTIDES

Peptides, and especially lipopeptides, have attracted recent attention as potentially active ingredients in skincare products, with the aim of improving the appearance of aged skin. These are sometimes termed 'cosmeceuticals', as they combine cosmetic properties with (purportedly) those of pharmaceutical actives. These agents are applied topically. There have now been a number of studies measuring the efficacy of these compounds, quantified through various techniques. These techniques include cell studies (assays of skin cells, i.e. fibroblasts, viability, and markers) and tests on tissue stiffness and hydration using animal skin or reconstituted skin models. In addition, qualitative and quantitative tests of skin appearance and wrinkling have been developed. None of this constitutes a recommendation of actual efficacy, however. As with other topical applications, these materials are not subject to rigorous testing as with pharmaceutical actives

Table 5.6 Examples of peptides used in skincare products.

Peptide[a] (common and trade names)	Source	Type	Mode of action
C_{16}-KTTKS (Matrixyl™ Palmitoyl pentapeptide-3 or -4)	Derived from the pro-collagen 1 sequence	Signal peptide	Stimulates production of collagen and other extracellular matrix proteins
GHK(Cu) and C_{16}-GHK, GHK(Cu) is copper tripeptide-1 or Lamin™	GHK is a metal ion-chelating sequence	Signal peptide/carrier peptide	Stimulates collagen and GAG production
FVAPFP or D-amino acid version, Peptamide™ 6	Extract of yeast fermentation	Signal peptide	Stimulates collagen production and up-regulates growth factors, matrix proteins, and others
N-acetyl-QNVH, acetyl tetrapeptide 9, Dermican™	Targeting of lumican metabolism	Signal peptide	Stimulates production of collagen I and lumican
GPRPA, pentapeptide 3, Vialox™	Snake venom	Neurotransmitter inhibitor peptide	Antagonist of the acetylcholine receptor leading to muscle relaxation (Botox alternative)
Acetyl-EEMQRR-NH_2, Argireline™	Derived from the SNARE protein SNAP-25, a substrate of botulinum toxin (Botox)	Neurotransmitter inhibitor peptide	Inhibits neuro-transmitter (catecholamine) release leading to muscle relaxation (Botox alternative)
Soybean peptides (mixtures with 3–6 residues)	Soy beans	Enzyme inhibitors	Inhibit proteases, increase collagen synthesis and decrease the number of apoptotic cells

[a] C_{16} denotes hexadecyl or palmitoyl.

such as the peptide drugs discussed elsewhere in this chapter. Table 5.6 provides examples of peptides used in skincare products. A diversity of sequences and quite distinct targets and modes of activity are apparent. Lipidation is performed in order to improve permeability/compatibility with the lipid membranes in the cells in the epidermis. Lipidation has also been shown to lead to self-assembly into fibrillar structures in the case of C_{16}-KTTKS and C_{16}-GHK, for example. Signal peptides are designed to elicit a biological response whereas carrier peptides deliver an active.

BIBLIOGRAPHY

Al Musaimi, O., Al Shaer, D., de La Torre, B.G., and Albericio, F. (2018). 2017 FDA Peptide Harvest. *Pharmaceuticals* 11: 10.

Badman, M.K. and Flier, J.S. (2005). The gut and energy balance: visceral allies in the obesity wars. *Science* 307: 1909–1914.

Castelletto, V., Hamley, I.W., Whitehouse, C. et al. (2013). Self-assembly of palmitoyl lipopeptides used in skin care products. *Langmuir* 29: 9149–9155.

Castelletto, V., Kirkham, S., Hamley, I.W. et al. (2016). Self-assembly of the toll-like receptor agonist macrophage-activating lipopeptide MALP-2 and of its constituent peptide. *Biomacromolecules* 17: 631–640.

Clemmensen, C., Muller, T.D., Woods, S.C. et al. (2017). Gut-brain cross-talk in metabolic control. *Cell* 168: 758–774.

Cobb, S.L. and Sit, C.S. (2015). Anti-infective peptides. In: *Advances in the Discovery and Development of Peptide Therapeutics* (eds. G. Kruger and F. Albericio), 97–110. London: Future Science.

Demmer, O., Just, R., Rottländer, M., and Fosgerau, K. (2015). Peptides in metabolic diseases. In: *Advances in the Discovery and Development of Peptide Therapeutics* (eds. G. Kruger and F. Albericio), 113–130. London: Future Science.

Gorouhi, F. and Maibach, H.I. (2009). Role of topical peptides in preventing or treating aged skin. *International Journal of Cosmetic Science* 31: 327–345.

Hamley, I.W. (2017). Small bioactive peptides for biomaterials design and therapeutics. *Chemical Reviews* 17: 14015–14041.

Hamley, I.W., Kirkham, S., Dehsorkhi, A. et al. (2014). Toll-like receptor agonist lipopeptides self-assemble into distinct nanostructures. *Chemical Communications* 50: 15948–15951.

Hamman, J.H. and Steenehamp, J.H. (2011). Oral peptide drug delivery: strategies to overcome challenges. In: *Peptide Drug Discovery and Development: Translational Research in Academic and Industry* (eds. M. Castanho and N.C. Santos), 71–90. Weinheim: Wiley-VCH.

Henninot, A., Collins, J.C., and Nuss, J.M. (2018). The current state of peptide drug discovery: back to the future? *Journal of Medicinal Chemistry* 61: 1382–1414.

Hutchinson, J.A., Burholt, S., and Hamley, I.W. (2017). Peptide hormones and lipopeptides: from self-assembly to therapeutic applications. *Journal of Peptide Science* 23: 82–94.

Hutton, J.C. and Siddle, K. (1990). *Peptide Hormone Secretion. A Practical Approach*. Oxford: Oxford University Press.

Jones, R.R., Castelletto, V., Connon, C.J., and Hamley, I.W. (2013). Collagen stimulating effect of peptide amphiphile C_{16}–KTTKS on human fibroblasts. *Molecular Pharmaceutics* 10: 1063–1069.

Kastin, A.J. (2013). *Handbook of Biologically Active Peptides*. New York: Academic Press.

Lau, J.L. and Dunn, M.K. (2018). Therapeutic peptides: historical perspectives, current development trends, and future directions. *Bioorganic & Medicinal Chemistry* 26: 2700–2707.

Lewis, A.L. and Richard, J. (2015). Challenges in the delivery of peptide drugs: an industry perspective. *Therapeutic Delivery* 6: 149–163.

Miller, M., Ambegaokar, K.H., Power, A.E., and Kaumaya, P.T.P. (2015). Peptides and cancer: vaccines and immunotherapy. In: *Advances in the Discovery and Development of Peptide Therapeutics* (eds. G. Kruger and F. Albericio), 85–95. London: Future Science.

Neal, J.M. (2016). *How the Endocrine System Works*. Chichester: Wiley Blackwell.

Negahdaripour, M., Owji, H., Eslami, M. et al. (2019). Selected application of peptide molecules as pharmaceutical agents and in cosmeceuticals. *Expert Opinion on Biological Therapy* 13: 1275–1287.

Nguyen, J.-T. and Kiso, Y. (2015). Delivery of peptide drugs. In: *Peptide Chemistry and Drug Design* (ed. B.M. Dunn), 157–201. New York: Wiley.

Nielsen, P. (2004). *Pseudo-Peptides in Drug Discovery*. Weinheim, Germany: Wiley-VCH.

Norman, A.W. and Henry, H.L. (2014). *Hormones*. Amsterdam: Academic Press.

Sunna, A., Care, A., and Bergquist, P.L. (eds.) (2017). *Peptides and Peptide-based Biomaterials and their Biomedical Applications*. Cham, Switzerland: Springer.

Tager, H.S. and Steiner, D.F. (1974). Peptide hormones. *Annual Review of Biochemistry* 43: 509–538.

de la Torre, B.G. and Albericio, F. (2017). The pharmaceutical industry in 2016. An analysis of FDA drug approvals from a perspective of the molecule type. *Molecules* 22: 6.

World Health Organization (2019). WHO Model List of Essential Medicines.

Wu, L. (2019). Regulatory considerations for peptide therapeutics. In: *Peptide Therapeutics: Strategy and Tactics for Chemistry, Manufacturing and Controls* (ed. V. Srivastava), 1–30. Cambridge: Royal Society of Chemistry.

Yu, M.Z., Benjamin, M.M., Srinivasan, S. et al. (2018). Battle of GLP-1 delivery technologies. *Advanced Drug Delivery Reviews* 130: 113–130.

Index

Introduction to Peptide Science, First Edition. Ian W. Hamley.
© 2020 John Wiley & Sons Ltd. Published 2020 by John Wiley & Sons Ltd.